MODERN INDUSTRIAL CITIES

SAGE FOCUS EDITIONS

MODERN INDUSTRIAL CITIES
HISTORY, POLICY, AND SURVIVAL

EDITED BY
BRUCE M. STAVE

 SAGE PUBLICATIONS   Beverly Hills   London

For information address:

SAGE Publications, Inc.
275 South Beverly Drive
Beverly Hills, California 90212

SAGE Publications Ltd
28 Banner Street
London EC1Y 8QE, England

Printed in the United States of America

Library of Congress Cataloging in Publication Data

Main entry under title:

Modern industrial cities.

    (Sage focus editions ; 44)
    Papers presented at the Conference on the Dynamics of Modern Industrial Cities, Comparative Perspectives on Order and Disorder held at the University of Connecticut in Storrs in the fall of 1979.
    Bibliography: p.
    1. Cities and towns—Congresses. 2. Cities and towns—History— Congresses. 3. Housing—History—Congresses. 4. Urban economics—Congresses. 5. Urban policy—Congresses. I. Stave, Bruce M. II. Conference on the Dynamics of Modern Industrial Cities, Comparative Perspectives on Order and Disorder (1979 : University of Connecticut)
HT119.M6   307.7'6   81-14497
ISBN 0-0839-1760-0   AACR2
ISBN 0-8039-1761-9 (Pbk.)

FIRST PRINTING

Portions of this volume first appeared in the *Journal of Urban History* and *Transactions of the Institute of British Geographers*.

For B.R.S.,
*whose fathering I admire
and enjoy increasingly
with each passing year*

CONTENTS

ACKNOWLEDGMENTS

In organizing the conference on The Dynamics of Modern Industrial Cities: Comparative Perspectives on Order and Disorder, which served as the origin of this volume, I was aided greatly by those of my colleagues at the University of Connecticut who served on the Program Committee: Robert Asher, Richard D. Brown, A. William Hoglund, Lawrence Langer, and Emiliana P. Noether. Emiliana Noether was especially helpful with translations into English, and Dick Brown, while serving as Department Chairman, first suggested that such a meeting be held. His advice and support were particularly appreciated.

Kathleen Madden, Claire Goodrich, and Jean Manter helped in innumerable ways to make order from what all too frequently seemed to be the disorder mentioned in the conference's subtitle. Janet Nolan, a doctoral student at the University of Connecticut, assisted with the preparation of this volume for publication and deserves thanks for her intelligent approach to the task. Figures for the essays by Kenneth Jackson and Brian Berry were prepared by Marilyn Richardson. Marjorie Ellsworth helped with some final typing, and Sharon Youland served beyond the call of duty when reading proof.

The funding for the conference, which was provided by the Rockefeller Foundation, the University of Connecticut Research Foundation, and the Aetna Life and Casualty Foundation, is greatly appreciated. The meeting was held in cooperation with the U.S. Department of Housing and Urban Development. The University of Connecticut Research Foundation also generously assisted in bringing this volume to publication.

The essays by François Bédarida and Anthony Sutcliffe and by John Modell were originally published in a special August 1980 issue of the *Journal of Urban History,* edited by me. Bédarida and Sutcliffe's work is based on some of the results of a research project financed by the Direction de l'Architecture, Ministère de la Culture, Paris. Modell's research was supported by the Philadelphia Social History Project, Theodore Hershberg, Director. His essay benefited from Hershberg's comments as well as from those of David Hogan, Michael Katz, Elaine T. and Larry May, and Russell Menard. The PSHP gratefully acknowledges the funding it receives from the following federal agencies: the Center for Studies of Metropolitan Problems, National Institute of Mental Health (MH 16621); the Sociology Program, Division of Social Sciences, National Science Foundation (SOC76-20069); the Division of Research Grants, National Endowment for the Humanities (RO 32485-78-1612); and the Center for Population Research, National Institute for Child Health and Human Development (RO 1 HD 12413).

A modified version of Kenneth Jackson's essay appeared in the same issue of the *Journal of Urban History.* An earlier version of the essay, which appears in this volume, was presented as the first Letitia Woods Brown Memorial Lecture to the Columbia Historical Society and the George Washington University in January 1977. Jackson wishes to thank, for comments on that draft, Herbert J. Gans, Robert Kolodny, William E. Leuchtenburg, and John A. Garraty of Columbia University; Joseph B. Howerton, Jerry N. Hess, and Jerome Finster of the National Archives; Frederick J. Eggers, Mary A. Grey, and William A. Rolfe of the Department of Housing and Urban Development; Joan Gilbert of Yale University; Mark Gelfand of Boston College; Joel A. Tarr of Carnegie-Mellon University; Margaret Kurth Weinberg of the Connecticut Governor's Office; and Christine Von Seggern of Chappaqua, New York.

Brian Berry's essay was delivered at the Conference on the Dynamics of Modern Industrial Cities, but was first published in the *Transactions of the Institute of British Geographers*, Volume 5, Number 1, 1980. It is published in this volume with the permission of the Institute. Berry wishes to thank William Alonso, Franklin James, Larry Long, Ira Lowry, Peter Morrison, George Peterson, and David Segal for their insights incorporated in his work. Stephan Jonas's essay was translated from French by Brigitte Fichet and Emmy Jonas.

The insights of the many unnamed participants in the conference, often incorporated in the introductions to the four parts of this volume, greatly enhanced all of our understanding of the issues relating to the dynamics of modern industrial cities.

Finally, my wife, Sondra Astor Stave, and my son, Channing, continued to support me in my many endeavors. During the period leading up to the conference, Sandye took time away from directing a program, Hartford: The City and the Region, to assist me in ways too numerous to count, as she did while I prepared this volume for publication. Channing, in keeping with his fascination for numbers, became intrigued by 1980 census population trends at the same time that he batted .368 in the Coventry Baseball League. Both deserve my gratitude for putting up with my preoccupations—urban and otherwise.

—BMS

PREFACE

In the year that the United States commemorated its bicentennial, Sam Bass Warner, Jr., suggested that urbanists may have less to celebrate than those marking the nation's two hundredth birthday. He pointed out that social scientists and planners frequently pay little attention to the historical roots of the processes and events that concern their specialities. On the other hand, urban historians increasingly tend to be limited by the narrow confines of the case study. As a consequence, Warner contended that planners proceed without much attention to scholarship, while scholars give little heed to the needs of planners or to the research findings and insights of other disciplines.[1]

With this in mind and in an effort to bridge the gap, in the fall of 1979, urbanists from across the United States and Europe were invited to the University of Connecticut in Storrs to participate in the conference on The Dynamics of Modern Industrial Cities: Comparative Perspectives on Order and Disorder. Historians, sociologists, planners, economists, political scientists, geographers, and policy makers gathered to discuss many of the essential issues of urban history and to consider its potential as a policy science.

The conference was divided into four sessions, which correspond to the four parts of this volume: The Role of Family and Neighborhood; Class Tension and the Mechanisms of Social Control: The Housing Experience; The Economy of Cities; and The Survival of Industrial Cities. Each session featured two speakers, whose papers had been distributed in advance, and two commentators. Themes were treated from an American and European perspective, with English, French,

German, and Italian scholars appearing on the program. The final session, consisting of a panel of commentators, explored general questions related to the survival of industrial cities and attempted to offer cohesion to the two days of discussion, which was spirited and stimulating. The introductions to the four parts of this volume summarize many of the issues raised by conference participants during the discussion periods following each session; they also aim to highlight the essential concerns of the authors and their critics.

As the essays within should indicate, many issues of interest to urban historians receive attention. While this volume may be diverse in geographic and temporal span, ranging from the streets of nineteenth-century London and Paris to the inner cities of contemporary America, certain unifying themes emerge. Consideration of the influence of ecological and cultural factors, of the private market, and of government policy on urban development all help to move us toward a better understanding of the dynamics of modern industrial cities. The nature of social control and the role of reformers in the urban context equally are matters of significance for those who study the development of such cities.

However, admittedly, the comparative perspectives on order and disorder suggested by the conference's subtitle did not always clearly emerge, although Part II of this volume comes closest to dealing with such matters. Moreover, the potential of urban history as a policy science still remains to be determined. Not long after the fall foliage that greeted conference participants turned to compost, a critic of the meeting wrote, "One suspects, given the easy and unselfconscious coupling of urban history and 'policy science,' that some notion of applied history danced in their [the conference organizers'] heads, but what is obvious is that the links between urban history and policy are very weak, and cannot be strengthened without major intellectual advances. One suspects, though, that the implicit guiding idea behind the planning of the conference was that the future of urban history did lie in its application to policy, and that the strategy of the meeting was to mingle historians with planners in the hope of stimulating policy-minded history and historically minded policy.''[2] While the latter suspicion is accurate and the notion of applied history did dance in the editor's head like visions of sugar plums at Christmas, the assertion that the links between urban history and policy are weak may be as dogmatic as a claim that urban history is a policy science. The jury is still out and we await a decision.

There is wisdom in David J. Rothman and Stanton Wheeler's assertion, when considering social history and social policy, that they could not conclude that history had no significant contribution to make to social policy. Their investigation, however, made them keenly aware of the barriers to cooperation between historians and decision makers and the difficulties of bringing them together.[3] Like Rothman and Wheeler, I acknowledge the difficulty in linking history and policy, but also recognize the promise. If one expects to make immediate decisions based on the essays included in this volume, one will probably be most disappointed. However, pondering the function of the street even in cities and times gone by, investigating family educational choices made many years ago, considering government housing policy made a half century before today or the thinking of housing reformers many decades before that, may be of value to those who make important decisions that shape our society.

It would be incorrect to believe that such individuals do not have a sense of history. For example, in a study of two major urban policy debates of the past twenty years, Seymour Mandelbaum finds that history, if not the historian, plays an important role in policy formulation. Far from caring too little about history, the policy community considers it important enough to define and interpret in its own and, perhaps, self-serving way.[4] As a consequence, Eric Lampard's suggestion that the best use of history may be as an anti-policy science takes on added meaning. While asserting that urban history could contribute to the education of "our masters—namely those experts and functionaries who will make our urban critical choices," Lampard has suggested that the primary responsibility of urban historians is to the public, the laity, or readers. The task of some historians, at least, is not to confirm but to be critical of the policy makers and of their policy scientists. Historians should hold up allegedly "'scientific' findings and rhetoric, especially when they contain generalizations about the past, to broader standards of historical experience." In one respect, the job of the historian is to shake the certainty of the policy maker by introducing more varied and less systematic considerations. "The more historical experience we can persuade policy science to include in the parameters of its models, the more likely its predictions [will] end up as tentative and confused as the exogenous world it presumes to straighten out," Lampard has stated.[5]

The role of the historian, then, may well be to show to the *real world* what the *real world* is. If the link between urban history and policy is dependent on "major intellectual advances," what might such advances be? Three recent articles on U.S. urban history have suggested that progress of analysis rests with increasing application of Marxist interpretation.[6] While the impact of such interpretation on U.S. urban history is still to be determined, its absence is recognized. In his recent consideration of contemporary historical writing in the United States, Michael Kammen notes that despite the relative gains made by Marxist scholars in America, "there are important fields such as urban history and the 'new political history' where Marxist influence has been relatively slight."[7] This may be attributable to a heavy ethnocultural emphasis in both areas or to reasons more arcane.

Nevertheless, the dialogue between European and American scholars regarding the dynamics of modern industrial cities sometimes demonstrated different vocabularies stemming from an uneven understanding of Marxism on the part of both groups. Whether such interpretation "is the looming frontier in urban analysis"[8] is also still to be determined, as is the question of whether the link between urban history and policy will be firmly welded by movement in such a direction. The essays and commentary in this volume may offer no definitive answer to

these questions, but they may serve as scouts on the way to this
new frontier. At the least, it is hoped that they will neither stand
as false guideposts nor lead to dead ends.

—BMS

Coventry, Connecticut

NOTES

1. Sam Bass Warner, Jr., and Sylvia Fleisch, "The Past of Today's Present," *Journal of Urban History* (November 1976), 3-4.

2. James E. Cronin, "The Problem with Urban History: Reflections on a Recent Meeting," *Urbanism Past & Present* (Winter 1979-1980), 42-43.

3. David J. Rothman and Stanton Wheeler, eds., *Social History and Social Policy* (New York, 1981), 2-3.

4. Seymour J. Mandelbaum, "Urban Pasts and Urban Policies," *Journal of Urban History* (August, 1980), 454.

5. Eric E. Lampard in Bruce M. Stave, *The Making of Urban History* (Beverly Hills, CA, 1977), 283-284.

6. Cronin, "The Problem..."; Michael Frisch, "American Urban History as an Example of Recent Historiography," *History and Theory* 18 (1979); Michael H. Ebner, "Urban History: Retrospect and Prospect," *Journal of American History* (June 1981), 69-84.

7. Michael Kammen, ed., The Past Before Us: Contemporary Historical Writing in the United States (Ithaca, NY, and London, 1980), 25.

8 Cronin, "The Problem...," 43.

I

THE ROLE OF FAMILY AND NEIGHBORHOOD

Family and neighborhood are among the most basic institutions that shape and are shaped by urban life. This section includes two essays, and commentary upon them, that approach these institutions from varying perspectives. François Bédarida and Anthony Sutcliffe consider the physical and social importance of streets in nineteenth-century Paris and London, while John Modell offers an ecological analysis of family decisions regarding suburbanization, schooling, and fertility in Philadelphia during the final two decades of the nineteenth century and the first two of the twentieth.

Whether the dynamics of modern industrial cities shape the streets of our metropolises or the streets shape what occurs within them is a question debated by scholars. Less controversial is Bédarida and Sutcliffe's assertion that the street merits serious historical attention. Certainly the inability of streets in the twentieth century to cope with the automobile is a policy issue of great significance. Moreover, understanding the relationship of private and public space is essential to comprehending the development of modern cities. Attitudes toward the street in the London and Paris under study differed as much as the perception that in Paris one might *stroll*, while in London one was considered to *loiter*. The development of grand boulevards as opposed to narrow thoroughfares reflected an amalgam of decisions, which are complex in motivation and intent. As David Goldfield noted in his conference commentary, one must consider a number of variables dealing with economics, demography, politics, culture, and technology when investigating differences in street development. The limits of ecology when studying street life are apparent in cross-national comparisons. For example, in southern European nations, street life is lively regardless of broad or narrow space; in Sweden, it is sparse. Hence, one might ask when considering Bédarida and Sutcliffe's work why French culture is more oriented toward street life than is English. Rather than standing as independent variable, space may be better understood as the dependent variable.

One conference participant, Giogio Piccinato of the Institute of Architecture of the University of Venice, suggested that Bédarida and Sutcliffe's research was based on a nostalgia for the "better" streets of the past. Inquiring why the street has ceased to be what it used to be, Piccinato asserted that modern streets are artificial. The nineteenth-century street claimed an important commercial role; it served as an instrument to condition consumers to buy products. Today, using the mass media, there are other means of conditioning. With the street no longer the only place to carry the message of the market, according to the Italian planner, modern street life has decayed. Another participant noted the absence in Bédarida and Sutcliffe's work of any consideration of the separation of children from adults in social space. Contending that children were the "leading colonists" of the street, he questioned research based on what Sutcliffe himself called "a sort of vague idea of the integration of all ages in the population."

If children have been among the chief users of streets for play and other activity, they have also seen their street life curtailed by the necessity in modern cities to attend school. John Modell found that the quality of schools during the period he studies, measured by expenditure per student, did not vary greatly between Philadelphia's peripheral city wards and those located closer to the city center. As a consequence, after World War I, families with a large number of children who sought quality education trekked to outlying politically independent suburbs, which were willing to spend more on instruction per student. Suburbanization was encouraged.

As Martin Katzman pointed out during the conference, land and housing space were often cheaper on the periphery, and this economic factor must be considered in any analysis of early twentieth-century suburbanization. Moreover, he raised these questions concerning Modell's study: Do suburban households care more about raising many and high-quality children than urban and rural families? If so, does the suburban milieu merely appeal to people who already have these tastes, or does environment have an effect? Are urban-suburban differences in familism simply the spatial expression of social class differences? In general, one must consider how behavior and environment interact to generate urban change and whether space acts as a receptacle or an activator. In such a consideration, other participants called for an investigation of child care, the nature of the female labor force, transportation, and ethnic and religious variation.

It would appear, then, that the role of family and neighborhood cannot be considered in isolation. Furthermore, as cost-conscious Americans move through the 1980s, Modell's measure of school quality—expenditure per student—comes increasingly into question with issues such as California's Proposition 13 and Massachusetts' Proposition 2½. Cities such as Boston suffer the brunt of the attack as advocates of economic conservatism and supply-side programs maintain that you cannot solve society's problems, let alone urban ills, by throwing money at them. Hence, contemporary policy makers, as well as scholars, would do well to evaluate Modell's premise. Consideration of the nineteenth-century streets of London and Paris may also have more than historical significance. Whether it is the responsibility of the historian to demonstrate this significance is another matter which divides the discipline.

1

THE STREET IN THE STRUCTURE AND LIFE OF THE CITY

Reflections on Nineteenth-Century London and Paris

FRANÇOIS BÉDARIDA
ANTHONY R. SUTCLIFFE

THE STREET AS A HISTORICAL PROBLEM

The street, it must be admitted, does not cry out for historical attention. The historical study of the use of physical space is still in its infancy, and the idea of splitting up the totality of space into distinct components has scarcely even been formulated, let alone justified. Urban historians, of course, assume that a valid distinction can be made between urban and rural space, but to distinguish between various elements of urban space is another matter. On the other hand, a partial breach in orthodox attitudes has already been secured by the history of housing. Although the spatial implications of housing history are rarely spelled out, it is implicitly assumed that it is useful to concentrate on the private

Authors' Note: *This chapter is based on some of the results of a research project financed by the Direction de l'Architecture, Ministère de la Culture, Paris. It was originally published in the Journal of Urban History, Vol. 6, No. 4, August 1980, 379-396.*

environment of the individual or household, to the virtual exclusion of the wider environment over which no direct personal control can be exercised. An important component of this wider environment is, in fact, the street. Outside the dwelling or its immediate, and privately controlled, curtilage, lies a shared space which must be traversed before the individual can enter other privately controlled spaces such as shops, factories, or the homes of friends and relatives. In venturing into this shared space, the individual implicitly accepts a public code of behavior which will almost always differ from his domestic conventions. It may be, of course, that precisely because the street is shared space, it is also neutral space, with its code of behavior merely a lowest common denominator, incapable of playing a creative role in social life. In today's city, after all, people spend very little time in the street.[1] Certainly, very little serious historical work has been produced on the street, apart from purely technical studies of street building and maintenance.[2] Mention ought, however, to be made of *De straat: vorm van samenleven* (Eindhoven: Stedlijk Van Abbe-museum, n.d. [1972]), an ambitious exhibition catalogue edited by Jjeerd Deelstra, Jan van Toom, and Jaap Bremer. This endeavors, very seriously and thoroughly, to isolate the historical experience of the street. Much of it, nevertheless, is very general urban history, as indeed is a more recent effort in this genre, Arlette Farge's *Vivre dans la rue à Paris au XVIIIe siècle* (Paris: Gallimard/Julliard, 1979).[3] Although none of this is particularly encouraging, let us try to look more closely at the street.

WHAT IS THE STREET?

The street can be defined basically as public circulation space in towns.[4] Like all brief definitions, this one immediately raises problems. Some circulation space, such as squares and parks, seems to surpass the street. At the other extreme, courts, alleys, closes, and mews seem to qualify only with difficulty, either because they do not admit wheeled traffic or because they are in private ownership. It is in the "immediate vicinity" of the dwelling that the problem of definition becomes most intractable. In London, for instance, many courts were, in the eyes of the law, part of the public street, even though they might have been entered by narrow alleys or tunnels. In Paris such courts were normally private, belonging to the owner of the building or buildings to which they gave access, but their function was clearly that of a shared space and they were in effect an extension of the street. And the problem does not end there, for in Paris most residents still had to pass through a staircase and/or corridor before reaching their own front doors. Is the staircase in an apartment block, then, also part of the street? For the purposes of this article, we shall crave tolerance for our decision to set aside the staircase, and the shared corridor, and the court, as intermediate areas, but we would not pretend that this expedient resolves the fundamental problem.

Let us return now to the indisputable street—the thoroughfare, normally lined by buildings, and wide enough to accommodate animals and wheeled traffic. Since Roman times at least, no Western European town has been able to exist without such streets. We can define a town as a community which has been liberated from direct dependence on primary production by the development of trading, manufacturing, administrative, or recreational functions. Thanks to this liberation, the town can grow, in both population and area, beyond the size of the agricultural village. However, in contrast to those of the village, the town's functions are nearly all spatially *intensive*—they depend on a high degree of accessibility and they thus tend to concentrate not only within the perimeter of the town but also at or around one or more nodes within the town. There results from this concentration the main physical distinction between town and country, the much greater density of building in the former.

The ultimate physical expression of these centralizing forces would, no doubt, be a solid mass of building. However, such an extreme degree of concentration would destroy the town.[5] Because the town depends on accessibility, the free movement of people and goods *within* the town must be assured at all costs. Thus, as soon as a town possesses buildings, it must also possess

streets to give access to them.[6] This simple reality is reflected in
many early definitions of "town." According to the *Encyclopédie,*
in 1765, a town is "un assemblage de plusieurs maisons ras-
semblées par rues et fermées d'une clôture commune."[7] Indeed, by
the eighteenth century, the interdependence of street and building
was clearly recognized; Expilly, for instance, wrote in his
Dictionnaire géographique "[on] crée des rues pour créer des
maisons." However, there has always been a limit to the area
which streets may occupy. In theory at least, they must be held to
the minimum consonant with the efficient exercise of the town's
functions, for they are not themselves productive space (and for
this reason are usually in communal ownership). Great expanses
of street, by reducing the accessibility of residents one to another,
would undermine the principal advantage of the town over the
countryside. We consequently find that the towns of medieval
Europe maintained networks of very narrow streets, for the prevailing methods of transport did not require anything more. In
the twelfth century, a law of Henry I required that an urban
thoroughfare (*straet* or *strata*) should be wide enough for two
carts to pass each other or for 16 armed knights to ride abreast.[8]
In reality, many streets were much narrower. In London there
existed numerous "lanes" (*venellae*), a term originally applied to a
rural track, which are referred to in the thirteenth and fourteenth
centuries as varying between five and twelve feet in width.[9] In
Paris many such narrow thoroughfares, often bearing the title of
"ruelle," survived into the nineteenth century and beyond.

Although we can assume that, at the time of the creation of
such narrow streets, an equilibrium existed between the spatial
demands of communications and those of other functions within
the town, there was a tendency for this equilibrium to break down
over time. If an efficient unit, the town would generate growth,
while the broader currents of economic development would tend
to produce means of transport requiring more space. Pressure on
the streets, or at least the more frequented of them, has thus been
a common feature of urban life throughout the ages. Of course,
the limits of the streets are not fixed for all time. They may be
widened or redirected by public authority. On the other hand,

private buildings may infiltrate onto the street. The physical arrangement of a town may therefore be regarded as the product of a struggle for space between (public) communication functions and other (private) uses.

This close relationship between public streets and private sites lies at the core of the street's *second* major function, as an architectural baseline. In northern Europe at least, it has been customary since the early middle ages for the streets to precede building, in that houses have normally been built on existing thoroughfares or new ones laid out deliberately to create building sites. This precedence has allowed the street to become the main organizing force in the townscape,[10] particularly as it has been normal ever since the middle ages for the principal buildings on any urban site to be erected directly on the street frontage. This practice, no doubt the product of considerations of accessibility and security, has produced the street's distinctive physical character, as a corridor formed by parallel lines of building frontages.[11]

It is clear from the long, narrow form of medieval urban plots that the areas away from the streets were originally cultivated. As more streets were built, to form distinct blocks, a tripartite arrangement of space emerged, consisting of the street, the frontage houses, and the area of enclosed, and private, open space within the block. In view of the urban model outlined above, it is understandable that the progress of economic development should lead to the progressive disappearance of these open spaces. On the one hand, existing spaces came to be covered by buildings and, on the other, newly created sites became shallower. This process of densification reached its conclusion in the nineteenth century when, notably in Paris, blocks were laid out without any interior space at all.[12] As the blocks became smaller, the role of the street as an architectural organizer was further enhanced. However, the use or disuse of this interior space within the block has been an important influence on the *third* main function of the street, that of providing a *social space*.

We have already seen that in the medieval town the areas occupied by streets are the product of a conflict in which they are reduced to the smallest possible area consonant with the efficient *economic* functioning of the town. The street is also almost the only public space in the town, if we exclude churches, the town walls, and other exceptional areas. The street thus becomes the locus of a whole range of interactive activities. Moreover, the street's social function is likely to become more important as interior spaces within the blocks disappear and residents, on leaving their dwellings, are increasingly forced into the street. To some extent, of course, pressure on the street can be reduced by the provision of special accommodation for activities of a commercial or even a social character, such as shops , market halls, and inns. However, until the nineteenth century, the level of economic development greatly restricted this provision, the use of which in any case normally required the individual to make some kind of payment. Consequently, the street, at any rate until the twentieth century, retained a multifunctonal character. The result, inevitably, was congestion as the street, its area determined by the minimum needs of communication, had to accommodate a growing volume of social and commercial activities. In the medieval town, only the large marketplace, which usually took the form of a broadening of the thoroughfare, indicated that the street's inability to accommodate a full range of functions had been recognized. And even such marketplaces often became choked with buildings during periods of municipal indifference to the public needs of the town.

These pressures reached their peak during the nineteenth century, when the traditional townscape had to cope with unprecedented growth. The result was the overloaded, kaleido-scopically multifunctional street lined by parallel, continuous facades of ever-increasing height, which Gustave Doré carica-tured in his London sketches around 1870.[13] However, it was precisely because industrialization appeared to offer other ways of organizing towns that this grossly overloaded street prompted social reformers and architects to consider removing some of its functions. The antistreet movement was to reach its ultimate development in Le Corbusier's dictum, formulated in the 1920s, of "il faut supprimer la rue." This nineteenth-century turning point is central to our inquiry.

THE NINETEENTH-CENTURY STREET

Although the nineteenth century put new pressures on the street, not least in the multiplication of forms of street transport, it also produced a much greater regulation of the street and its adjoining buildings. The authorities took increasingly effective action to pave, drain, light, and police it. Building codes related building heights closely to street widths, producing regular, constant-height facades. As building densities and heights increased,[14] the individuals increasingly became a prisoner of the street. His dwelling, if it had a view at all, would look onto the street. On leaving his home, he would quickly find himself in the street. In moving across the town he would proceed through a succession of streets, his eye constantly imprisoned in a single perspective without the lateral distraction of open spaces or interior courtyards. Encounters with individuals, street furniture, animals, and vehicles would require him to make constant changes of course, precluding relaxation. The street would be a scene of conflict, ultimately expressed in its role as a battleground in civil disorders.

On the other hand, the concentration of functions in the street made it an exciting and convenient place to be, and growing regulation made it much more comfortable in the nineteenth century than ever before. The spread of sanitary provision gradually reduced the danger of chamberpot drenchings, gutters and downpipes stopped rainwater sluicing into the street, and underground drains and rudimentary street cleansing allowed the pedestrian to keep a clean pair of legs, incidentally allowing trousers to replace breeches and stockings and precluding the need for thick soles and high heels. This new comfort of the street was promoted most dramatically by the introduction of the sidewalk, which had been an almost unknown luxury before

1800. The sidewalk, when combined with a convex carriageway, gutters, and underground drains, completely transformed the use of the street. Pedestrians could now walk along the sides of the street instead of in the middle, undisturbed by vehicles and animals, and sometimes shaded by trees. Kiosks, drinking fountains, and urinals could be provided on the sidewalks.

This canalization of the movements of a happier, more relaxed pedestrian along the building frontages encouraged shopkeepers to try to catch his eye with window displays. In Paris, cafe owners managed to establish their right to put tables and chairs on some of the sidewalks. In London, neither the climate nor the attitude of the authorities permitted such a use of public space, but in the poorer districts many shops used the sidewalks for the display of goods. This highly successful rearrangement of the street, complete in both London and Paris by the third quarter of the nineteenth century, helps to explain why the galleries, arcades, and passages, which had proliferated in both cities between about 1800 and 1850, faded thereafter. They were outshone by that apotheosis of the street, the Haussmannic boulevard, and even by its more modest London equivalents.

So if the individual was the prisoner of the nineteenth-century street, he was lulled by a number of compensations which, for a time, must have appeared to be improvements over previous arrangements. These improvements help to explain why the antistreet reaction was postponed until the twentieth century, when the arrival of the motor vehicle tipped the balance against the street. Indeed, even as late as the early 1900s, the influential Parisian architect, Eugène Hénard, the greatest traffic-engineering expert of his time, was still trying to retain the street by upgrading its capacity through the use of several levels of circulation and by easing the claustrophobic effect of the continuous facades through the introduction of setbacks (redans).[15]

Why, then, could the street not cope with the motor vehicle? The question brings us back to the close link between streets and buildings. The nineteenth-century renaissance of the street was associated with a dense, street-oriented habitat which was beginning to cause concern on health grounds when the motor vehicle suddenly created a new source of danger, and of noise and fumes which were trapped in the street. The motor vehicle also demanded a new series of street improvements which the growing densification of buildings along the thoroughfares made more costly than ever in the past. Although it required a massive mental effort to envisage a town without streets, once the ideal image had been created, notably by Le Corbusier, it proved to be very persuasive. Indeed, it is still with us today, in that multifunctional streets are rarely included in planning schemes for new districts.

THE STREET AS SOCIAL SPACE

We may well regret the loss of the social intensity which the nineteenth-century street could provide. However, no simple return to the past is possible. We have to recognize that use of the street and attitudes toward it are conditioned by the general state of social organization. For instances, the distinction between public and private space was much clearer in the nineteenth century than it had been previously, owing to the progress of individualism and the modern idea of intimacy, under the influence of bourgeois romanticism. Within the home, spaces were now clearly designated for distinct purposes, and the individual came to expect a greater degree of privacy. This more complica- ted, and therefore more expensive, lifestyle was permitted by an increase in the level of real incomes, and it consequently pursued its course during a nineteenth century in which incomes rose for all but the very poor. Clearly, the street could not meet this demand for the appropriation of space for personal use, and by the end of the nineteenth century it may well have been perceived as a bigger threat to the individual than it had been at the beginning. Sam Bass Warner has worked on the nineteenth- century effort to privatize the street, not so much by taking over the street itself, but by creating private spaces such as hotel lobbies or select suburbs from which undesirables are excluded, or choose to exclude themselves.[16] Insofar as they were forced to use the street, argues Warner, different social classes sought to protect themselves from intrusions by adopting a distinctive appearance, a point which has also been taken up by Richard Sennett in *The Fall of Public Man*. To the nineteenth-century rich, efforts to merge imperceptibly into the crowd merely invited unpleasant encounters; far better to adopt an aggressive style of dress which, though it might attract the occasional mendicant, would prompt a deferential response in almost everyone else.

During the twentieth century, continuing social change has further altered popular attitudes to the street. Private space is now more highly prized than ever, and clearly there can be no question of re-creating patterns of nineteenth-century street life today. On the other hand, there is no need to accept the current "reductionist" tendency which would dismiss the street as merely a useless relic of the past.[17]

THE STREETS OF LONDON AND PARIS[18]

We also have to be aware of contemporary variations in the role of the street. By the nineteenth century, Paris and London had come to present a number of divergences. Until the eighteenth century, Paris had been rather larger than London and had developed, according to the common Continental pattern, very high residential densities.[19] By the sixteenth century there were indications that an increase in building heights was being accompanied by a densification of building within the blocks. This extension of building away from the street raised the problem of access, for the tunnels and passages, through which court houses were normally approached, by no means offered the same convenience as the street, and the rents which these buildings could bear were consequently lower than those charged in comparable buildings on the street frontage. The most effective solution, which was to drive new streets across the existing blocks, was clearly out of the question, but it was perfectly possible to reduce the depth of blocks created by new streets on the outskirts. We find the first clear indications of such an enhanced orientation toward the street in the development of the site of the Hôtel Saint-Pol by three new streets in 1544. The streets were unusually wide (10 meters) and their rectilinear pattern was clearly intended to permit the division of the land into convenient, rectangular sites.[20] By the seventeenth century the evolution toward a site of broader frontage but shallower depth can no longer be questioned, and by the eighteenth century we begin to encounter square sites in new districts, combined with the devotion of an unprecedentedly high proportion of the total area to streets.[21] Contemporary land-value estimates confirm the picture, indicating a sharp fall in land values away from the street. In 1730, for instance, an inquiry suggested that beyond a distance of six *toises* (11.6 meters) from the street, values fell by between a quarter and a half.[22]

These tendencies help to explain the efforts of the royal authorities, which became increasingly effective, from the reign of Henri IV in the early seventeenth century, to secure a network of broad, rectilinear streets.[23] The most impressive of these were the "boulevards," broad thoroughfares lined by trees, the first of which were created on the line of the demolished right-bank fortifications in the 1670s. Complemented by another royal creation, the avenue of the Champs-Elysées which André Le Nôtre began in 1670, the boulevard or avenue became the paradigm of the major thoroughfare in Paris. The entries of royal roads into the city were laid out in that fashion in the eighteenth century, and an outer ring of boulevards was built from 1784. By 1789 there were some 30 kilometers of ring-boulevards in Paris, most of them 40 meters in width.

So firmly established was this tradition of wide thoroughfares that it survived the Revolution and went on to inspire the ambitious improvements of the Second Empire. Although the boulevards carried increasingly heavy loads of traffic, they were also valued for their social function. The original boulevards of the 1670s had been designed as a shady promenade for citizens escaping the overcrowded city, and the first carriageway had been a narrow strip between four rows of trees. Flimsily built places of entertainment and refreshment had sprung up on them and had survived the subsequent widening of the carriageway. In the nineteenth century, when the boulevards were lined by permanent buildings, these social functions occupied parts of them while still retaining a foothold in the street itself. By the 1820s the Boulevard des Italiens had replaced the covered galleries of the Palais-Royal as the social center of Paris, and a class of leisured men-about-town, the *bouldvardiers,* came into existence. This equation of street life with the height of elegance was a potent image, and the success of the northwestern section of the *grands boulevards* contributed greatly to the general renaissance of the Parisian street during the nineteenth century (see Figure 1.1).

Paris began to grow rapidly again in the nineteenth century after near-stagnation in the eighteenth, and the street network expanded substantially. However, high-density apartment living remained the norm even in the outer districts. Consequently, most of the city, apart from a narrow suburban fringe, continued to be composed of streets lined by continuous facades of tall buildings. Increasingly, a distinction emerged between main traffic streets, on which commerce also concentrated, and narrower side-streets which were mainly residential in function, but a general intermixture of uses survived, with the upper floors being used for residence, even in the most commercial districts, until the later part of the century. This multiplicity of functions and high density sustained a varied street life throughout much of the city. The homes of the majority of the population were exiguous and, with few interior spaces available within the blocks, people seeking to escape their dwellings were thrown immediately onto the street. This distinctive balance between private and public space produced the phenomenon of the "flâneur," or stroller, an individual virtually unknown in London where French visitors remarked on the unnerving way in which pedestrians hurried along with scowls on their faces. There was

Figure 1.1: The Building Exteriors of the Boulevards des Italiens: top, about 1850; bottom, about 1885

even a synonym for flâneur, "badaud," a term which has almost disappeared from French usage today. In English, one may "stroll" in a park, but hardly in a street. On the contrary, "loiter," which Harrap's gives as a translation of "flâneur," has a distinctly sinister meaning in English.

This rich animation of the Parisian street was captured well by J. Letaconnoux on the eve of World War I:

La rue est un des cadres essentiels de la vie parisienne. Avec son animation, son bruit, ses personnages familiers, ses célébrités, ses passants pressés et badauds, avec ses boutiques et ses cafés, ses théâtres de boulevard et ses amuseurs ambulants, elle est une des curiosités, un des charmes de Paris; c'est un spectacle continu, divers et gratuit.[24]

The social attraction of the Parisian street was enhanced by the regular, impressive character of the architecture which a combination of high land values and strict building regulations engendered. Although opinions vary on its aesthetic qualities,[25] it was greatly respected by Parisians and tourists alike. Even if the pedestrian were imprisoned in the Parisian street, everything combined to make his confinement a pleasant one.

In London a more complex network of streets emerged from a more scattered pattern of building and the lack of any powerful regulation authority. In the nineteenth century, in particular, many existing villages were incorporated into the built-up area while, despite the efforts of the turnpike trusts, few wide rectilinear routes emerged. There was no system of ring roads comparable to the boulevards in Paris. The central streets were as crowded and as densely built as those of Paris, but on leaving the City one soon encountered thoroughfares of a recognizably suburban character. When outward growth accelerated from midcentury, extensive districts of even lower densities were built, producing mile after mile of two-story villas and terraces, often fronted by gardens. Few streets of a Parisian scale ever emerged. Regent Street, built by John Nash in the 1820s, significantly enough at royal command and partly on royal land, surpassed in grandeur anything in Paris at that time, but it could never be repeated. On the contrary, the distinctiveness of London streets lay in their variety. This variety achieved even a linguistic expression. In Paris, only a handful of generic terms were applied to thoroughfares (*rue, avenue, boulevard,* and, much more rarely *cours*). In London, at least *60* terms were in use in the later nineteenth century (e.g., street, road, way, place, terrace, row, avenue, broadway, highway, lane, grove, drive, and so on). Many of them were in fact rural survivals and were applied to suburban residential streets (see Figure 1.2).

In suburban London, the street offered little excitement or distraction. Dwellings were generally larger and more comfortable than their Parisian equivalents, and most had private

Figure 1.2: Regent Street about 1890. This photograph brings out architectural perspective, since the cupola that tops the corner building is reminiscent of the Paris boulevards.

yards or gardens. The street was a place of social intercourse and recreation only in the poorest districts, and the resulting form of low street life could scarcely appeal to more prosperous social groups or to tourists. On the contrary, the police made efforts to suppress or at least to control it, equating it with a criminal subculture which, in some parts of London, it certainly was.[26] And the weather constantly ruined any attractions which the street might otherwise have held out to the "stroller," even in the spaciously laid-out "great estates" of the West End.

CONCLUSION

We have argued above that the street—more spacious, better organized, more comfortable and exciting, and more important as social space—played a much bigger role in the life of nineteenth-century Paris than it did in London. It is curious therefore to discover that the most influential anti-street planning prophet of the twentieth century, Le Corbusier, formulated his proposals in direct reaction to his experience of Paris.[27] In the Radiant City of 1930, Le Corbusier presented a reconstruction of Paris without streets, but retaining, or so he hoped, the social and aesthetic qualities of a city which had captivated as well as horrified him. Paris, too conservative and too crystallized to accept such a radical transformation, has largely been spared the erruption of the Corbusian townscape. However, many other cities, including London, have been more seriously affected. They have, in fact, suffered from the results of a master-architect's mental and spiritual struggle with a dominating and suffocating urban form.

London, on the other hand, never generated a reaction against its streets. The reaction, if any, was against its sprawling suburbs, and the result was the garden-city idea of Howard and later the green belt and new towns associated with Patrick Abercrombie. The loose, irregular arrangement of buildings along thoroughfares was not seriously questioned until the later 1940s, and even then the inspiration came from abroad.

In the 1970s, Paris has been swept by a new reaction, this time against the partial introduction of Corbusian methods into renewal schemes. The traditional Paris street has again been presented as the ideal structuring element for the city, and the building regulations have been modified to protect its physical form. Thus, having generated Le Corbusier, Paris now appears to have generated an equally violent return to tradition.[28] These clear statements of alternatives are quite alien to London, which continues to muddle through without taking up definite positions for or against the street. In this way we can see that the street, by playing different functional roles in the two cities, has also affected the quality and nature of their artistic and intellectual production. And on this ground, we rest our case that the street merits serious historical attention.

NOTES

1. Note the virtual disappearance of male headgear in recent years. Bare-headed males are very rare in nineteenth-century urban photographs, probably because much longer periods of the day were spent in the open, especially by workers.

2. See, for example, the recent article by Clay McShane, "Transforming the use of urban space: a look at the revolution in street pavements, 1880-1924," *Journal of Urban History*, 5 (May 1979), 279-307.

3. The road has attracted rather more attention than the street, and some of the studies devoted to it are of some urban relevance. See, for example, Hans Hitzer, *Die Strasse: vom Trampelpfad zur Autobahn, Lebensadern von der Urzeit bis heute* (Munich, 1971).

4. Cf. a British legal definition of 1889: "The expression *street* includes any highway or other public place, whether a thoroughfare or not" (An Act to abolish any duties on coals leviable by the Corporation of London, 52 and 53 Vict., c.17).

5. Frank Lloyd Wright's idea of a central business district composed of one giant tower, as yet unrealized, would in theory achieve this ultimate concentration. However, lateral movement would simply be replaced by vertical movement, with the lift shafts acting as streets.

6. See Bernard Rouleau, *Le tracé des rues de Paris: formation, typologie, fonctions* (Paris, 1967), 8.

7. *Encyclopédie ou Dictionnaire raisonné*, vol. 17, 1765, article "Ville."

8. Lady Stenton, *English Society in the Early Middle Ages*, p. 254.

9. See E. Ekwall, *Street-Names of the City of London* (Oxford, 1954).

10. Cf. T. Sharp, "The English tradition in the town: the street and the town," *Architectural Review*, 78 (468), November 1935, 179-187; F. Pick, "The Street," *Architectural Review*, 74 (445), December 1933, 215-219.

11. Cf. the Oxford English Dictionary's definition of "street": "A road in a town or village (comparatively wide, as opposed to a 'lane' or 'alley') between two lines of houses; usually including the sidewalk as well as the carriageway. Also the road together with the adjacent houses."

12. This process is the subject of a report in the same research program as our own: Jean Castex et al., *De l'îlot à la barre: contribution à une définition de l' architecture urbaine* (2 vols.) (Versailles: Association pour le développement de la recherche sur l'organisation spatiale [ADROS], 1975).

13. See Blanchard Jerrold, *London: A Pilgrimage* (1870).

14. The counter-tendency of suburbanization did not become significant over most of Europe until the very end of the nineteenth century.

15. See P. M. Wolf, *Eugène Hénard and the Beginning of Urbanism in Paris, 1900-1914* (Paris, 1968).

16. See Warner's forthcoming article, "The public invasion of private space and the private engrossment of public space in the American metropolis 1870-1970," in *Growth and Transformation of Cities* (Stockholm: National Swedish Council for Building Research).

17. See the vigorous critique of this "reductionism" in S. Anderson (ed.), *On Streets* (Cambridge, 1978), vii.

18. This project has been based upon a detailed comparison of three pair of streets: Regent Street/Boulevard des Italiens, Bethnal Green Road/Rue Saint-Antoine, and Rosebery Avenue/Boulevard Magenta.

19. See Rouleau, op. cit., passim.

20. Rouleau, op. cit., 61.

21. See Rouleau, op. cit., pp. 79-80. For an even fuller study of Parisian land development, see H. Dumolin, *Etudes de topographie parisienne* (3 vols.; 1929).

22. Quoted in Gaston Bardet, *Naissance et méconnaissance de l'urbanisme* (Paris, 1951), 383-384.

23. For the development of regulations affecting streets, see M. G. Jourdan, *Recueil de règlements concernant le service des alignements et de la police des constructions dans la Ville de Paris* (Paris: Préfecture de la Seine, 1900).

24. J. Letaconnoux, *La vie parisienne au XVIIIe siècle* (Paris, 1914), 75.

25. For a critique of nineteenth-century Parisian architecture, see A. Sutcliffe, "Architecture and civic design in nineteenth-century Paris," in *Growth and Transformation of Cities*, op. cit.

26. The equation of street life with criminality emerges strongly from H. Mayhew, *London Labour and the London Poor* (1861-1862).

27. This point cannot be fully argued here, but see A. Sutcliffe, "A vision of utopia: optimistic foundations of Le Corbusier's *doctrine d'urbanisme,*" in Russell Walden (ed.), *The Open Hand: Essays on Le Corbusier* (Boston, 1977), 216-243.

28. See Paris-Projet, no. 13/14, n.d. [1975].

2

AN ECOLOGY OF
FAMILY DECISIONS
Suburbanization, Schooling,
and Fertility in Philadelphia,
1880-1920

JOHN MODELL

Jumbled clichés reflect the frustration felt in 1916 by parents in the flourishing northwestern section of Philadelphia, as a group of them petitioned the Board of Public Education. "The children of to-day are the men and women of our Countries [sic] future greatness, and we ask you in the name of the Lord that we are not held backward in sending our full quota to take their places in the World's progress. Give us schools and we will send the children."[1] Rapid population growth in the urban periphery had outstripped school construction. School author-

Author's Note: *This chapter originally appeared in the Journal of Urban History, Vol. 6, No. 4, August 1980, 397-417.*

ities could predict demand clearly enough and tried to meet it, but finances were too tight.[2] "It would be a happy omen and an attractive prospect to those who contemplate moving into a section," wrote the Superintendent of Schools in 1909, "if they could see in advance of their coming signs on selected sites, reading 'Property of the Board of Public Education.'"[3] The wish remained unrealized.

The situation was deemed particularly acute in the high schools, but the lower schools were also thought to be divided unevenly between the central and peripheral areas of Philadelphia.[4] In 1910, for instance, the Board of Public Education conceded that at the urban fringes "the pupils do not have as good an opportunity for securing an education as they would have in many of the townships of the State in which centralized schools have been established to which the pupils are transported at public expense."[5] Four years later the Board President looked forward to the time, now not far distant, "when every child in Philadelphia will have an opportunity to secure an education that meets the standards of the best city schools." But this was adjudged still "utterly beyond the power of the Board" seven years later.[6]

If complaints were justified, however, it was not (as we shall see) that peripheral schooling was inferior, but that public schooling in general was poorly supported in Philadelphia. Not unlikely, many parents in the "suburban" wards[7] wanted particularly good education for their children and communicated this wish strongly to school officials. An emphasis on high-quality education is amply documented in many suburban places that grew after World War II,[8] and much in ecological theory of urban growth would lead us to anticipate it in the urban periphery at an earlier date. No less than in the later period, families in the "streetcar suburbanization" phase of metropolitan development might be expected to have moved to the city's fringes to find the separation from inner-city temptations and the space—inside and outside of the home—on which middle-class children were assumed to thrive. Residence at the urban periphery implied a reorientation to "familistic" activities and away from both commercialized entertainment located in the downtown and from the classic locus of bourgeois economic accumulation.[9] If—as was the case after World War II—"suburban" families tended to have (for their socioeconomic level) relatively many children as well, the elements were in place that would exert heavy pressure on school officials for more and better schooling opportunities in peripheral wards.[10]

The city is a consequential man-made environment, its form the product of diverse and variously powerful preferences. As metropolitan areas have grown, technological and political mechanisms have been found that have permitted the construction of a set of local environments (or at any rate a subset) ever more finely tuned (for better or for worse) to the preferences of their own inhabitants. Spatial units are politically represented in both formal and informal political arenas, so that local preferences are both product and producer of the location of activities in urban space. Institutions that provide services do so in particular spatial contexts, and demands upon them are place-specific.[11] Urban space is not just an empty container; it affects the formation of tastes, in children and in adults. Where local preferences are embedded in institutions, their impact can extend over generations. In the twentieth century schools have become the most important public aspect of childrearing. Their quality and their social and geographic inclusiveness have always been facts of importance to the lives of their inmates.

This article seeks to discuss trends and spatial patterns in three complexly interrelated urban phenomena, as they developed in Philadelphia over the period 1880-1920. The phenomena to be discussed are childbearing and childrearing; the supply of and the demand for schooling; and suburban-style residential growth at the periphery of the city. I will analyze the relationship among the three dimensions under study, paying special attention to the changing contexts of family decision making (about fertility, residence, and school enrollment of children) and to the possible contribution of political decisions about the provision of schooling to the choices adopted by families living in different parts of the city.

The period of "streetcar suburbanization" in Philadelphia was one of rapid change in some but not all of the dimensions I am discussing. On average, Philadelphia women moved toward younger marriages over the period, while at the same time the pace of fertility within marriage lessened—the two offsetting tendencies, together, pointing to the spread of birth-control practices. Childrearing changed markedly. Census data show that both boys and girls continued considerably longer in school by 1920 (three to four times as many attended school in their late teen years), while the contribution of coresident children to complex family economies declined correspondingly. Public schools—and presumably the competing private and parochial schools[12]—increased their per-enrollee expenses, partly as a reflection of the far higher instructional costs in secondary schools but partly reflecting a modestly growing commitment to higher quality education. These changes accompanied a trend toward suburbanization in Philadelphia as in most American cities. On the periphery, home ownership was relatively common, and in Philadelphia at the end of the period under study, peripheral growth encouraged an overall gain in home ownership. In most cities, this trend was associated with increased proportions of single-family housing, but in Philadelphia the row house had long predominated, and there was no gain in single-family housing, although in the 1910-1920 decade a strengthened correlation between the location of single-family dwellings and of home ownership suggested the emergence of large and distinctively "suburban" areas.[13]

In each of five distinctive land uses for which comparable data are available for 1880 and 1920, remarkable continuities of extent and kind of spatial differentiation appeared. Churches closely followed the spatial distribution of households in both years; banks were always quite independent of residential patterns. Halls became a bit more ubiquitous over the 40-year period, and factories a bit more concentrated, but the changes were not great. Yet, over the period, churches, banks, halls, and factories (but not shops) became relatively rarer in "suburban-housing" areas, as the suburbs became differentiated as more strictly residential elements of the urban ecology.[14] To choose to live in a "suburban" section of the city implied by 1920 a greater separation from "urban" facilities than before, at least by contrast to those living in central Philadelphia.

During this period, however suburbanization left a great deal of Philadelphia's ecology basically unaffected. Tables 2.1 and 2.2 indicate the nature of this continuity, as seen in the relationships among ward fertility level, "suburban" housing, socioeconomic level, nativity composition, and the ratio of factories to families.[15] Table 2.1, showing zero-order correlations, points to a fertility considerably reduced in areas of favored socioeconomic composition, and to a lesser extent in wards in which relatively few foreign families lived. Factories nearby bore little relationship to fertility, while "suburban" wards contained in each year slightly fewer child-prone families.[16] (The "suburbs," to be sure, were areas with many native whites whose fertility tended to be low.) One new and suggestive pattern in 1920 was a significantly negative relationship between "suburbs" and factory location, as already noted.

When statistical controls are imposed, in Table 2.2, we see that "suburbanization" has become more closely related to all the other variables by 1920. Relatively more of the 1880 urban ecology than that of 1920 had been explained by common ties with socioeconomic level; in 1920, elements of the built environment made more difference. The 1920 city saw "suburban" housing more closely articulated to the socioeconomic dimensions, and even more apparently to the absence of industrial activity. Wealthier areas at both dates were associated with lower marital fertility, a relationship that was enhanced by taking into consideration the housing context in which families lived. This pattern intensified over the 40-year period. At the same time (holding ward socioeconomic level statistically constant), families living in areas of "suburban" housing may be seen to have had more children, another relationship that strengthened over the period. Thus, housing context was complexly related to fertility

TABLE 2.1   Pearsonian Correlations among Five Selected Ward
Characteristics, 1880 and 1920

			1880	
	Fertility	"Suburbanization"	Socio-economic Level	Nativity Composition
"Suburbanization"	−.260	x	x	x
Socioeconomic level	−.591***	.002	x	x
Nativity composition	−.132	.715***	.289	x
Factory ratio	−.129	−.071	−.146	−.051

			1920	
	Fertility	"Suburbanization"	Socio-economic Level	Nativity Composition
"Suburbanization"	−.304*	x	x	x
Socioeconomic level	−.543***	.245*	x	x
Nativity composition	−.217	.635***	.657***	x
Factory Ratio	.179	−.424***	−.219	−.191

NOTE: Sources and calculation of indicators differ from 1880 to 1920, as explained.
   *Significance, .05
  **Significance, .01
***Significance, .001

(or to the residential choice of families according to their prior
fertility); directly, through the increasingly strong tie of the
suburbs with fertile families, and indirectly, through the suburbs'
strengthened overlap with prosperous areas, which were themselves increasingly associated with low-fertility families. Such
major family decisions as number of children to have were, thus,
becoming more closely articulated with the residential grain of
the city, precisely the kind of association that could, over time,
promote neighborhoods homogeneous in taste for family living,
and thereby encourage lasting changes in tastes. But at the same
time, through 1920, countervailing tendencies kept these relationships below the surface.[17]

TABLE 2.2   First-Order Correlations of Five Selected Ward Characteristics with Ward "Suburbanization" and Socioeconomic Level, Controlling for Three Ward Characteristics, 1880 and 1920

CORRELATIONS WITH "SUBURBANIZATION," CONTROLLING FOR:	Fertility	"Suburbanization"	Socio-economic Level	Nativity Composition	Factory Ratio
			1880		
Socioeconomic level	.323*	x	x	.747***	−.071
Nativity composition	.511**	x	−.307*	x	−.049
Factory ratio	.254	x	−.009	.714	x
			1920		
Socioeconomic level	.536***	x	x	.649***	−.391**
Nativity composition	.586***	x	−.296*	x	−.399**
Factory ratio	−.426***	x	.172	.623***	x
CORRELATIONS WITH SOCIOECONOMIC LEVEL, CONTROLLING FOR:					
			1880		
"Suburbanization"	−.612***	x	x	.412*	−.146
Nativity composition	−.538***	−.307*	x	x	−.137
Factory ratio	−.622***	−.009	x	.285	x
			1920		
"Suburbanization"	−.668***	x	x	.669***	−.131
Nativity composition	−.543***	−.296*	x	x	−.127
Factory ratio	−.524***	.172	x	.642***	x

*Significance .05
**Significance .01
***Significance .001

Table 2.3 explores the growing significance of the suburban context to the relationships under study, by examining these relationships separately in "suburban" wards and central-city wards for 1880 and 1920. By 1920 we find a weakened negative relationship between socioeconomic level and fertility in the "suburbs." This is historically significant, for in 1880 it had been the "suburban" wards where the closer relationship had obtained.[18] In the 1920 "suburbs," not socioeconomic level but the presence or absence of factories best explained fertility variation. Strictly residential "suburban" wards, in which few factory jobs

TABLE 2.3   Correlations with Fertility Measures in the Most "Suburban" Wards of Philadelphia, and in the Balance of the City, 1880 and 1920

	1880	
	In the "Suburban" Wards (N = 11)	In the Balance of the City (N = 20)
Correlation of fertility and:		
Socioeconomic level	−.786**	−.554**
Nativity composition	−.417	−.502*
Factory ratio	−.346	−.008

	1920	
	In the "Suburban" Wards (N = 16)	In the Balance of the City (N = 32)
Correlation of fertility and:		
Socioeconomic level	−.333	−.728***
Nativity composition	−.101	−.438**
Factory ratio	.527**	.206

*Significance .05
**Significance .01
***Significance .001

were available, were especially likely to be populated by families which chose to have few children, and which did not expect them to work as adolescents.

School enrollment in effect sums a pair of decisions—that of government or other institutions to provide opportunities (and perhaps the compulsion) for schooling and also of families to forego current income from children to take advantage of these opportunities. Throughout the period under study, the Board of Public Education of Philadelphia provided free universal public schooling through grade 8 and free high schooling being offered to a more or less select group of grade-school graduates. Rather than measure the supply of schooling (all historical measures of which depend upon counts of those who accepted schooling opportunities), I will examine the *quality* of the schooling Philadelphia provided to those who showed up at different locations. The logic of this procedure is that while school administrators would trim or augment particular teaching staffs in response to the numbers who agreed to attend, the per-enrollee resources they provided suggests the generosity of their response to parental demand. Thus, in 1880, wards where each teacher was on average better paid (men teachers, and more senior teachers, both thought to be better, were better paid) were also the wards where each teacher taught fewer students. These two elements, compounded, can be expressed as teaching expenditures per enrolled student—and it is this that I shall employ as my measure of schooling "quality." (The 1920 measure included all instructional expenses, which were mainly teachers' salaries.)[19]

In 1880, "better-quality" public schools were provided to wards that paid the highest per capita taxes to the city. This relationship (not a linear one) was "rational" perhaps, but not entirely egalitarian. There resulted a significant linear relationship between the socioeconomic level of ward residents and the "quality" of public schooling provided there. Yet even this relationship, though theoretically important, indicates that the spatial distribution of higher status residents explained only about 15 percent of the spatial inequality of school "quality." In part, this was because the "best-off" residents did not always live where the city collected most of its taxes. The tax-base bias of school "quality" favored center city, where land values were highest and in which nonresidential land uses had already pushed out families. "Suburban" wards, therefore, received "lesser quality" education, on average, although the relationship was not significant. Overall, the net effect of Philadelphia public schooling was to spread school expenditures relatively more evenly across the city's wards than tax money was collected.

Did the expansion of education over the next 40 years, and the several administrative reforms that intervened, change these basic relationships? The extent of spatial variation in school "quality" diminished only marginally between 1880 and 1920—indeed, identical "regional" patterns prevailed. As in 1880, there was a tax-base bias, and a small bias toward more prosperous areas. Once again, "suburban" wards and modestly "*lower*-quality" public schools, and in the "suburbs," schools were slower to put in place extensive kindergarten programs and classes in industrial and mechanical arts, sewing, and homemaking.[20] As in 1880, the effect of citywide school administration in Philadelphia was to redistribute school funds toward less well-endowed areas. Thus, while school "quality" varied across Philadelphia wards in 1920 about as much as it had 40 years earlier, variation among local school systems in nearby counties was over twice as great.[21]

The ecology of school enrollment (unlike fertility) has been little studied, especially for the prewar era. Professional educators concerned themselves with grade retardation and dropouts associated with "problems of foreign parentage and population congestion," and absent in "outlying residential and semisuburban sections of the city."[22] Table 2.4 indicates that in both 1880 and 1920, Philadelphia families living in prosperous and largely native wards kept their children in school significantly longer.[23] Ward enrollment levels were somewhat depressed in factory areas, but the "suburban" housing qualities of wards seemingly had little to do with enrollment. The relationship between school enrollment and fertility was sharply negative. Public school "quality" (not shown) was not significantly related to enrollment levels.[24]

The first-order and joint correlation coefficients between fertility and school enrollment levels by ward, in combination with each of the other independent variables, shown in Table 2.5, point again to the great degree of continuity of Philadelphia's urban ecology from 1880 to 1920. Note especially the uniform strengthening of the joint correlations: Families seemingly traded off between many and more-educated children in 1920 as in 1880, and their choices were apparently more conditioned than before by the nature of the area in which they resided.[25] One distinct change in the ecology of family choice does appear,

TABLE 2.4   Correlations of Ward Level of Extended Schooling with Selected Ward Characteristics, 1880 and 1920

	Correlation with Extended Schooling in:	
	1880	*1920*
Fertility	−.442**	−.695***
"Suburbanization"	.189	−.001
Socioeconomic level	.689***	.776***
Nativity composition	.468**	.397**
Factory ratio	−.327*	−.235

*Significance .05
**Significance .01
***Significance .001

however. In 1880, the strong negative correlation between fertility and extended schooling disappeared under statistical control for socioeconomic level; by 1920, however, school prolongation per se retained a highly significant negative tie with fertility level, even when socioeconomic level was controlled. Neighborhoods had become more sharply characterized by the choices local families made between many and more-educated children.

In 1920, as was not so much the case in 1880, this choice did not mainly flow from socioeconomic level. Since the intervening period had been one of increasing diffusion of effective contraceptive methods and of attitudes conducive to their employment, this trend is to be expected, since the notion of a trade-off between many and well-educated children rests on *articulated* family decisions. Urban transportation, contraceptive technology, a shorter workday, and a larger paycheck gave more Philadelphians the ability to make this kind of choice. Urban families by 1920 increasingly could and did craft their families and chose their residences in conjunction with their taste in family living.

The proximity of factories was a less significant context for family decisions about schooling in 1920 than it had been in 1880.

TABLE 2.5   First-Order and Multiple Correlations of Ward Level of
             Extended Schooling with Selected Ward Characteristics,
             1880 and 1920

	1880			
	"Suburbanization"	*Socio-economic Level*	*Nativity Composition*	*Factory Ratio*
First-order correlations of ward level of schooling with fertility, controls as in top stub	.518**	−.069	−.434**	−.516**
Joint correlations, with fertility and characteristics in top stub	.542*	.681***	.605**	.588**

	1920			
	"Suburbanization"	*Socio-economic Level*	*Nativity Composition*	*Factory Ratio*
First-order correlations of ward level of schooling with fertility, controls as in top stub	.730***	−.518***	−.680***	−.683**
Joint correlations, with fertility and characteristics in top stub	.730***	.842***	.739***	.704***

*Significance .05
**Significance .01
***Significance .001

In the earlier part of the period, factory jobs, which offered
relatively high wages to juvenile beginners, were critical in the
process by which children left school and entered a phase of
somewhat irregular employment, a shifting from job to job, and
often prolonged periods of nonemployment.[26] At school-leaving
ages, nonfactory jobs were to a good extent fill-ins where factory
work was not available.[27] In areas where factory jobs (for both

boys and girls) were available, children left school relatively young; this was not the case where available jobs were in shops or offices.[28] As factory work for children declined in the twentieth century, school enrollment "leveled up" across the city. Annual school censuses (arrayed in Table 2.6) indicate how closely the schooling history of successive cohorts was tied to nearby industrial employment. When the Depression of 1921 effectively destroyed the wartime boom in Philadelphia manufactures, factory jobs for juveniles disappeared, whereupon families modified their economies, permanently, as it turned out.[29]

The 1880-1920 period in Philadelphia, then, saw at least three developments of direct relevance to the origins of school-centered neighborhoods: (1) Wards became increasingly clearly differentiated according to the way residents chose between many and well-educated children. (2) In a variety of ways, the centrality of ward socioeconomic level to the urban ecology lessened over the period, although it was still important. (3) "Suburban" residential areas became more distinctive in their characteristics, notably in the near-absence of factories, and this dimension in a number of ways became a focus of Philadelphia's ecology. Since the 1910s was a decade of heightened suburbanization within Philadelphia, we may well suspect that at this point "suburban" neighborhoods emerged that catered to families who wished many *and* well-educated children.

Examination of zero- and first-order ecological correlations between fertility and school enrollment in "suburban" and in other wards, for 1880 and 1920, gives some comfort to this hunch—but only a little. In both 1880 and in 1920 (somewhat more so in 1920), the tightness of the trade-off of fertility and schooling was less in the "suburbs" than in center city; with socioeconomic level controlled, the trade-off in the suburbs was weak in 1880 and absent in 1920.[30] Ward-by-ward inspection of the data indicates that there were indeed a few "suburban" wards by 1920—but only a very few—in which the considerable economic resources of resident families were as usual associated with relatively extended school enrollment but in which fertility was not markedly low. These were, in fact, the very wards prone to

TABLE 2.6 Proportions of Residents Age 15 Enrolled at School, and Proportions of Residents Age 14-15 Employed in Factories, Two Selected Districts and Philadelphia, 1915-1922

	% Enrolled at Age 15			% Working in Factories at Ages 14-15		
	District 3	District 6	Philadelphia	District 3	District 6	Philadelphia
1915	37.6	40.4	48.5	20.8	21.5	16.9
1916	44.9	47.3	53.9	17.7	14.8	15.0
1917	61.4	57.0	62.0	7.8	7.8	11.7
1918	67.1	72.6	65.2	5.9	6.9	11.9
1919	74.5	73.1	64.4	4.2	8.1	10.6
1920	73.5	73.1	67.4	5.9	11.3	12.0
1921	77.2	71.2	71.8	3.0	8.3	6.0
1922	84.1	79.6	81.4	4.4	6.2	7.8

petitioning the public school authorities for increased facilities. Here, families were more prone to own their own homes than in the typical "suburban" ward, but not especially devoted to single-family housing; they were about average among "suburban" wards in nativity composition, but considerably on the high side socioeconomically. Each of these "familistic" wards was notably lacking in factories, even by "suburban" standards: Schooling was the "only game in town" for teenagers, as families surely had known in making their residential decisions. Schooling "quality," however, although on the high side *for the "suburbs,"* varied from slightly below the city average to slightly above it. The public school system gave no particular comfort to the emergent tendency toward "familism."

Within Philadelphia, no "familistic" political program—such as the Board of Public Education was seemingly asked to provide—influenced the city's ecology. But in the politically independent suburbs, where public schooling expenditures could vary far more widely, enclaves of rather high fertility combined with extended school attendance could be found. In these towns, instructional expenses per student were high; here, school "quality" had a strong and apparent relationship to the emerging "familistic" rejection of a trade-off between many and well-educated children. This new pattern was, as we have seen, largely absent within Philadelphia proper.[31]

In the postwar era, a new ecology was to emerge, its roots in earlier developments but its pace and its mechanisms new. Wartime savings were spent in part on new peripheral housing and on automobiles to get there. The locus of peripheral growth shifted decisively outside the central city, into politically independent suburbs. Incomes increased and families grew larger in both suburb and center city, but faster in the suburb, so that suburb/city contrasts were intensified, in wealth, large families, younger families, families with children, and families in the "early family life-cycle stage."[32] Children's contributions to family economies declined, and the positive relationship of child labor to fertility weakened. Wives worked more often, and the negative tie of wives' work to fertility strengthened. Prior fertility became less relevant to extended schooling.

The new ecology met a rapidly changing occupational universe in which college education was widely seen as a distinctly well-rewarded investment, which, in turn, focused the attention of local school systems upon their college-preparatory programs. A consensus endorsing such arrangements was particularly hearty in suburban areas that had been built up and inhabited in a brief time, where high proportions of families would share not only socioeconomic level but also stage in the family life cycle.[33] In the earlier period, we have seen, residential options did not really include this kind of suburb, at least not for that majority whose "suburb" was within Philadelphia proper. Where one lived, to be sure, was then as subsequently highly relevant to fertility and to investment in schooling. Between 1880 and 1920, local fertility levels seem to have enlarged their codeterminant relationship with local school enrollment levels, but contextual modifications of this relationship were, as we have seen, emergent. Local pressures, however, did little to produce a politics promoting new kinds of educational environments. It was only the new demography, ecology, and metropolitan political and educational structure of the postwar period that encouraged the proliferation of self-consciously school-centered areas.

In this sense, not only did postwar suburbanization more than ever before reflect a "family" dimension of social differentiation. Its very political auspices seem to have been agents of this process. By the same token, the implicit choice of the Philadelphia Board of Public Education not to discriminate in favor of exigent "suburban" wards in earlier years may be said to have had a lasting impact upon the social composition of the city, and perhaps upon the choice of some Philadelphians to move beyond its political boundaries.

NOTES

1. Richard H. Fisher, William F. Dixon, and Charles C. Cox to the Board of Public Education, March 8, 1916, School District of Philadelphia, Records Center. A systematic

examination of the largely unorganized School District Archives would surely shed light on the subject of this article.

2. A brief survey of the fiscal and derived difficulties of the school system in Philadelphia is presented in Thomas E. Finegan, "Oustanding Features of the Survey of the Schools of Philadelphia," in Pennsylvania, State Department of Public Instruction, *Report of the Survey of the Public Schools of Philadelphia* (Philadelphia, 1922), Book I, 11-30; Book II, Part I, Chapter 2. William W. Cutler III, "A Preliminary Look at the Schoolhouse: The Philadelphia Story, 1870-1920," *Urban Education* 8 (1974), 381-399. In 1913, almost 19,000 Philadelphia school children were on part time, and 1300 more (most likely in kindergarten) were awaiting admission altogether. Philadelphia, Board of Public Education, *Annual Report,* 1913, 15.

3. Philadelphia, Board of Public Education, *Annual Report,* 1909, p. 83. Over the 18-year period 1911-1929, the 17 wards of Philadelphia that were net losers of elementary schools gave up 46 sites, gained 16, and kept 63; 13 wards were net gainers, and these lost 20, gained 41, and kept 56; another 18 wards balanced losses with gains (13 sites), while keeping 61. School District of Philadelphia, Board of Public Education, *School Site and Construction Record 1911-1929* (1930).

4. Philadelphia, Board of Public Education, *Annual Report,* 1909, 12-23, 86-87; Pennsylvania, Superintendent of Public Education, *Report,* 1912, 260. Similar pressure was felt within the parochial school system. Archdiocese of Philadelphia, Superintendent of Parish Schools, *Twenty-Fifth Annual Report, 1919* (Philadelphia, 1919), 12-13.

5. Philadelphia, Board of Public Education, *Annual Report,* 1910, 714.

6. Philadelphia, Board of Public Education, *Annual Report,* 1913, 17-18; 1920, 17.

7. Inevitably, a certain degree of confusion is engendered by discussing the "suburbs" as elements of the city proper. But we must. The *process* of suburbanization, save for those elements traceable to political independence, was a long-term process that worked its way out toward and then beyond the periphery of the city. In Philadelphia, a massive consolidation in 1854 put areas within the political boundaries of the city that were ecologically suburban a century later. (At the same time, consolidation simplified our task somewhat by precluding annexations over the entire period we treat, although it was being discussed again in the 1910s.) In Philadelphia in 1880-1920, suburbanization was more characteristic of the political city than of areas surrounding it.

8. John R. Seeley, R. Alexander Sim, and Elizabeth W. Loosely, *Crestwood Heights: A Study of the Culture of Suburban Life* (New York, 1956), ch. 8; Robert C. Wood, *Suburbia: Its People and Their Politics* (Boston, 1959), 186-194.

9. E. Gartly Jaco and Ivan Belknap, "Is a New Family Form Emerging in the Urban Fringe?" *American Sociological Review,* 18 (1953), 551-557; Wendell Bell, "Social Choice, Life Styles, and Suburban Residence," in William M. Dobriner, ed., *The Suburban Community* (New York, 1958), 225-247; William Michelson, *Man and His Urban Environment* (Reading, MA, 1970); John Modell, "Suburbanization and Change in the American Family," *Journal of Interdisciplinary History,* 9 (1979), 621-646.

10. In the 1910s, Philadelphia's vestigial "ward boards" still were charged with receiving and transmitting citizen petitions. "Suburban" residents were by far the most active in petitioning their boards for enlarged school facilities, despite the presence in the most exigent of these of higher ratios of public (as also private) schools to population than in the city as a whole. Philadelphia, Board of Public Education, *Journals,* 1911-1916; William H. Issel, "Modernization in Philadelphia School Reform, 1882-1905," *Pennsyl-*

vania Magazine of History and Biography, 94 (1970), 358-383; Pennsylvania State Department of Public Education, *Report of the Survey of the Public Schools of Philadelphia,* Book II, Part I, Ch. 1.

11. Norton E. Long, *The Unwalled City: Rebuilding the Urban Community* (New York, 1972); John R. Logan, "Industrialization and the Stratification of Cities in Suburban Regions," *American Journal of Sociology,* 82 (1976), 333-348; Logan, "Growth, Politics, and the Stratification of Places," *ibid.,* 83 (1978), 404-416; David Harvey, *Social Justice and the City* (Baltimore: The Johns Hopkins University Press, 1973); Harvey, *Society, the City and the Space-Economy of Urbanism* (Commission on College Geography Resource Paper 18; Washington, DC, Association of American Geographers, 1972); Patrick J. Ashton, "The Political Economy of Suburban Development," in William K. Tabb and Larry Sawers, eds., *Marxism and the Metropolis* (New York, 1978), 64-89.

12. School censuses, available from 1904, reveal that the proportion of Philadelphia students outside the public school system remained basically stable. In 1920, for example, 24 percent of Philadelphia school children attended parochial school, and three percent attended private school. One would be naive to take the school censuses at face value, especially as regards enrollment rates. The Bureau of Compulsory Education, which took the census, had few canvassers, and they were not professional; their instructions did not call for inquiry at houses not known to have children living there. Since their inquiry was conducted in order to enforce compulsory education, one would hardly expect universal cooperation. Internal and external evidence, however, points to many substantial uses for which these data were adequate. Except for the 1904 census, published in Philadelphia, Board of Public Education, *Report of the Chief of the Bureau of Compulsory Education,* 1905, 9-13, the censuses are reported in the *Annual Reports* of the Bureau of Public Education.

13. Methodological discussion of the use of ward-level data for ecological analysis is found in Modell, "Suburbanization and Change in the American Family."

14. "List of Dwellings, Factories, Horse-Power, & c., as Charged on the Registers of 1880," Report of the Department for Supplying the City With Water, Appendix to Philadelphia, Select Council, *Journal,* 1880; "Tabulated Statement of Properties in the City of Philadelphia, Compiled from Assessors' Books of the Year 1920." Philadelphia City Archives. The "suburban" index for 1920 consists of an equally weighted addition of the standard (z-) scores of proportion of houses owner occupied and (reversed) ratio of households to dwellings. The owner-occupancy item was not available in 1880, and so different items indicating "suburban" qualities were indexed in like manner for that year. In addition to the households/dwelling ratio, tabulated from the manuscript census footings, the index includes a measure of the three-year growth rate of housing and of the absence of old fashioned "part" dwellings, derived from the Water Department data.

15. The 1880 fertility measure is an age-standardized marital fertility rate, based on coresident living children summed over each ward, and based on family-level data derived from the 1880 files of the Philadelphia Social History Project, themselves based on the manuscript census. The 1920 measure is more nearly unconventional: It is an index of the proportion of all legitimate births for 1919-1921 that were *not* first births, for each ward, to native-born and foreign-born women separately (so as to standardize for nativity composition of each ward). Fourth and subsequent births were negatively correlated with first births, as proportions of all births; non-first births correlated positively with more

conventional measures of birth rates. The 1880 socioeconomic-level indicator is the proportion of employed adult male residents who held white-collar jobs, derived from the project individual-level data. The 1920 measure is the average rental value of housing *for 1930,* from the U.S. Census of that year (Vol. VI). No adequate socioeconomic level measure for 1920 was available, since both the real property and personality tabulations available by ward pertain to undiscoverable degrees of business holdings. Use of the 1930 figure for 1920 assumes that the socioeconomic composition of wards does not change radically in a decade, a tenable assumption with a very few exceptions. Nativity composition is simple: percentage of adult males U.S.-born in 1880 (from Project individual level-data), and the percentage of population U.S. born of U.S. parentage (from the published census) in 1920. The factory ratio relates numbers of factories to households, based on sources cited in Note 14.

16. Prewar urban studies include: Philip Morris Hauser, "Differential Fertility, Mortality, and Net Reproduction in Chicago: 1930," unpublished Ph.D. dissertation, University of Chicago, 1938; Henry Allen Bullock, "Spatial Aspect of Differential Birth Rate," *American Journal of Sociology,* 49 (1943), 149-155; Warren S. Thompson, "Some Factors Influencing the Ratios of Children to Women in American Cities, 1930," *American Journal of Sociology,* 45 (1939), 183-199; Warren S. Thompson and Ruth J. Nelle, "Ratios of Children to Women in Chicago and Cleveland Census Tracts, 1930," *American Sociological Review,* 4 (1939), 773-791.

17. This conclusion holds true, we must recall, *on the average,* and *on the ward level.* Not impossibly, smaller, more homogeneous areal units would have revealed a closer tie between suburbanization and large families; and, as we shall see, certain whole wards deviated by 1920 from the average tendency, for reasons specific to themselves.

18. Several demographers before World War II began to observe slight tendencies toward an emergent positive relationship of fertility and economic resources, particularly in areas with "suburban" housing characteristics. Thompson and Nelle, "Ratio of Children to Women," 782; P. K. Whelpton and Clyde V. Kiser, "Social and Psychological Factors Affecting Fertility, I," *Milbank Memorial Fund Quarterly,* 21 (1943), 236-238, 245.

19. Until 1905, school expenditures by ward were published in the *Annual Reports* of the Board of Public Education, but thereafter by larger districts whose boundaries cut across wards. In 1921, however, data were published by schools in a way that could be aggregated up to wards, in "Financial Report of the Secretary," in Board of Public Education, *Annual Report,* 1921, Statement D, following page 338. My measure for school "quality" in 1921 included *all* instructional expenses in its denominator.

20. For some years before 1915, the Superintendent of Buildings published an appendix to his annual report tabulating various characteristics of school buildings put up within the last half dozen years. Examining per-pupil costs of new buildings constructed in the four "demanding" wards detailed in Table 2.1, in comparison with those built elsewhere, I found no differences.

21. The statistic is the coefficient of variation, .092 for Philadelphia wards in 1920, but .250 for all incorporated minor civil divisions in Delaware and Montgomery Counties, Pennsylvania, as recorded in Pennsylvania, Department of Public Instruction, *Statistics of the Public Schools 1919-20, 1920-21,* first part, Tables 3 and 7.

22. Board of Public Education of Philadelphia, Division of Educational Research, *Age Grade Survey* (Bulletin Number 146, February 1930); Byron A. Phillips, "Retarda-

tion in the Elementary Schools of Philadelphia," *The Psychological Clinic,* 6 (May-June, 1912), 1-6. But see J. B. Maller, "Vital Indices and Their Relation to Psychological and Social Factors," *Human Biology,* 5 (1933), 94-121.

23. Ward school-enrollment levels for 1880 are measured by an index of family school-proneness, summed over the ward. It is based on the proportions of children in attendance at school during the year at ages 11-12, 13-14, and 15-16, weighted according to citywide probability. The source of the original data is the Project family files. In 1920, I take advantage of the publication in the U.S. Census of proportions of children 16-17 attending school, by ward.

24. The matter of non-public schooling calls for considerably closer examination. School census data on 8 school "districts" from the 1920 school census indicate, for both boys and girls at school-leaving ages, that public schools in areas where instructional costs per student were greater (and class sizes smaller) enrolled smaller proportions of local children, while private schools enrolled relatively many. (Parochial schools showed a mixed pattern.) The area in which this preference for private schooling was most pronounced was, once again, in the northwest section of the city where, it seems, demand for fine education outran the modest efforts of the Board of Public Education to meet it.

25. Theoretically and empirically relevant perspectives are found in Richard Easterlin, "The Economics and Sociology of Fertility: A Synthesis," in Charles Tilly, ed., *Historical Studies of Changing Fertility* (Princeton, 1978), 57-134; Peter H. Lindert, *Fertility and Scarcity in America* (Princeton, 1978), ch. 6; Carl F. Kaestle and Maris A. Vinovskis, "From Fireside to Factory: School Entry and School Leaving in Nineteenth-Century Massachusetts," in Tamara K. Hareven, ed., *Transitions: The family and Life Course in Historical Perspective* (New York, 1978), pp. 135-185; John Modell, "Patterns of Consumption, Acculturation, and Family Income Strategies in Nineteenth-Century America," in Tamara K. Hareven and Maris A. Vinovskis, eds., *Family and Population in Nineteenth-Century America* (Princeton, 1978), 206-240.

26. Philadelphia, Board of Public Education, *Annual Report,* 1918, 159-160; 1911, 64.

27. James S. Hiatt, "The Child, The School, and The Job," Public Education Association [of Philadelphia] Study 39, 1912, 6, 9-10.

28. These conclusions are based on ecological analysis of the earliest published school census in which ward breakdowns are present, combined with assorted U.S. Census and local tax data.

29. Districts 3 and 6 (which had largely but not perfectly constant boundaries in the years shown in Table 6) were low-enrollment immigrant areas, with many textile and clothing factories. Comparable data are not available before 1915. Employment of children rose slightly between 1925 and 1930, when the Great Depression dropped employment of 14- and 15-year-olds by half within two years. By 1938, children under 16 were legally forbidden to leave school for gainful employment.

30. The correlation coefficient of school enrollment at 14-15 (the oldest narrow age group) and fertility (children 0-6/family) was slightly lower (-.540) than in the "suburban" wards of Philadelphia. If we confine ourselves to the 14 civil divisions outside of Philadelphia that grew more than 25 percent in the 1910s—suggestive of suburban development—the correlation coefficient was even less pronounced (-.470), suggesting even more play there for a "familistic" choice of *both* many and well-educated children.

31. As cited in Note 21 and U.S. Census for 1920.

32. Amos Hawley, *The Changing Shape of Metropolitan America* (New York, 1956); Otis D. Duncan and Albert J. Reiss, *Social Characteristics of Urban and Rural Areas,*

1950 (New York, 1956), 117-133; Bernard Lazerwitz, "Metropolitan Community Residential Belts, 1950 and 1956," *American Sociological Review,* 25 (1960), 245-252; Modell, "Suburbanization and change in the American Family"; and see Albert Hunter, *Symbolic Communities* (Chicago, 1974), which seems to indicate (pp. 30-33) that in Chicago, extra-city suburbanization was accompanied by a weakening of within-city differentiation by family characteristics.

33. Ernest R. Mowrer, "Sequential and Class Variables of the Family in the Suburban Area," *Social Forces,* 40 (1961), 107-112. For postwar school expenditure patterns in the Philadelphia area see Oliver P. Williams et al., *Suburban Differences and Metropolitan Policies: A Philadelphia Story* (Philadelphia, 1965). The data in Williams et al. suggest that the more suburban the milieu, the more its government tended to be influenced in its level of school support by "social rank" of its residents. The authors argue (p. 17) persuasively that "differentiation and specialization within metropolitan areas" had by 1960 produced "divergent local interests and policies that perpetuate[d] demands for autonomy."

3

COMMENTS

Martin T. Katzman:

The chapters in this section deal with the general issue of how diverse populations accommodate to each other in urban space. Modell's "An Ecology of Family Decisions" unifies four converging themes in an exciting extension of the traditional field of urban ecology: First, the cliometric revolution is spreading from economic events to other realms of social inquiry. Social scientists are increasingly concerned with quantitative tests of their propositions using historical data. Second, sociologists and economists are increasingly concerned about decision-making processes in households. The study of Philadelphia focuses on several outcomes: the number of children to have, whether or not females participate in the labor force, the quantity and quality of schooling children will receive, and the nature of the social class and ethnic environment in which the family will reside. Third, and inherent in the previous complex of decisions, is the choice of residential location, a perennial concern or urban sociologists, geographers, and economists. Fourth, political scientists, economists, and historians have increasingly focused on the means by which similar families satisfy collective wants in a heterogeneous society. As rightly indicated in Modell's chapter, schooling is the most important of these collective wants, and the one that may display the greatest spatial variation.

As indicated, the choices about family size, female labor force participation, quantity and quality of schooling, social milieu, and location are reciprocally, if not simultaneously, determined. Economists tend to view choices as a reflection of both opportunities and preferences, at least in the short run. Over the longer run the distinction between opportunities and preferences may be less tenable.

John Modell focuses on the changing relationship between *familism* and the *suburban milieu*. Familism refers to an orientation toward urban living, exemplified by rearing highly educated children. While suburbanization is ostensibly peripheral development at low densities, Modell endows the suburban milieu with other mystical properties. The evolving preferences toward familism and the growing relationship between familism and suburbanization are studied over the period 1880-1920.

Familism and suburban milieu are somewhat troublesome analytical concepts. Do suburban households care more about raising many and high-quality children than urban and rural families? If so, does the suburban milieu merely appeal to people who already have these tastes, or does this milieu somehow beatify those fortunate enough to approach its environs? Are urban-suburban differences in familism merely the spatial expression of social class differences? Is low housing density, social class homogeneity, or middle-class control of the schools the Holy Grail that was sought in suburbanization, or all of the above?

There are at least two reasons families with children are more likely to reside in the peripheral areas than families without: housing costs and schooling. The simple economic explanation, which Modell does not mention, is that land and housing space are cheaper on the periphery. Large families presumably demand more space than small families. The larger the family, the greater the savings in housing costs to be realized by moving outward an extra mile. This same outward propensity would benefit large households no matter what their social class or income level. Indeed, there was no zero-order correlation between suburbanization and social class in 1880 (Table 2.4). When social class is held constant, there is a positive relationship between suburbanization and fertility (Table 2.5).

For the middle class, another key intervening variable linking familism and suburbanization is schooling. Both folk pedagogy and modern educational research recognize that school quality depends on the peer environment, school resources like teachers and materials, and the cultural norms the school system inculcates. Let us consider each of these in turn.

Middle-class families can seek a relatively supportive peer environment for their children by self-segregation in the housing market. Conceivably, a homogeneous environment could be sought in either city or suburb. Philadelpha in the late 1880s was a destination for immigrants from Ireland, Germany, Italy, and other non-Protestant countries. Unfortunately, neither ethnicity nor race are given much consideration in Modell's account. Although social class was not highly related to nativity in 1880 (Tables 2.4 and 2.5), the Anglo-Protestant middle-class apparently sought ethnic homogeneity through suburbanization.

Since the cultural norms of the school are controlled more centrally through the selection of principals and teachers, the middle-class may have more difficulty in exercising its will in this realm. As Banfield notes, the largely native middle-class ceded control over big-city government to the largely immigrant working class by the end of the nineteenth century. Rather than achieving its will through domination of the central administration, the middle-class would have to rely on favoritism from the school board, or, failing that, secede from the larger polity.

The middle class was able to elicit only minor favoritism in the allocation of school funds. As a matter of policy, the school district allocated resources in proportion to property-tax collections. This rule does not necessarily favor the affluent, who have expensive homes, because a substantial proportion of the tax base is in commercial buildings and factories. These latter structures are more common in low-class wards (Table 2.4). In both 1880 and 1920, only 15 percent of the variance in school quality (as measured by expenditures per pupil) was related to neighborhood social class; moreover, school quality in the peripheral areas was about the same or slightly lower than in the center. In other words, through suburbanization the middle class could purchase space at low cost and provide a middle-class native milieu, but could not acquire superior school resources. While in Philadelphia the coefficient of variation in school expenditures was only .09, the variation was far greater (.25) among independent townships surrounding the city. This suggests that higher expenditures could be obtained by moving beyond the city boundaries, and thus seceding from the larger polity. Philadelphia at the turn of the century was relatively egalitarian in its school expenditures, for the coefficients for Atlanta, Boston, and Chicago in the 1960s were around .14.

Attendance figures also suggest that Philadelphia was a fairly egalitarian society. Attendance was rather weakly correlated with peer social class, own social class, and own family size ($R^2 = .14$). In contrast, educational attainment in the 1960s nationwide was much more closely related to own SES ($R^2 = .25$ from Jencks). Holding these factors constant, neither ethnicity nor school quality appears to matter. In other words, biases in the distribution of school resources in favor of the middle class apparently had little impact on their likelihood of attendance. Nor did Anglo-Protestant homogeneity provide advantage. The suburban milieu may have had some independent impact, because the availability of local industrial opportunities had a depressing effect on attendance.

It is unfortunate that the politically independent suburbs were excluded from the study. In 1880, the housing market and labor market under consideration was probably entirely within the confines of Philadelphia's political boundaries. By 1920, this was probably not the case. As the automobile opened up rural townships to suburban development, a key political decision was made at the state level to forbid further annexations by Philadelphia. A common pattern in the industrial states, this prohibition serves the interests of the Anglo-Protestant middle class quite well. It permits them to satisfy their collective wants for schooling. As Modell indicated, variations in school spending among independent suburbs were relatively high, and suburban spending was one-third higher than urban spending.

The conclusion of the Philadelphia study offers profound lessons. To paraphrase Modell, the choice of the Philadelphia Board of Education not to discriminate in favor of the familistic, middle-class wards had a lasting impact on the social composition of the city, because superior options developed beyond the city limit by 1920. The Anglo-Protestant policy of containment of immigrants within the city limits resulted in fiscal institutions conducive to middle-class flight. Two alternative institutions would have reduced this flight: (1) discriminating in favor of the middle class within the city, an unthinkable option in today's constitutional environment, or (2) annexing the suburbs and foreclosing the possibility of fiscal autonomy. It is titillating to note that liberal annexation laws are most common in areas where few immigrants lived when fiscal institutions were established (e.g., the Southeast and the Southwest). It would be instructive to see whether cities in these regions have undergone the same pattern of middle-class flight as Philadelphia.

The Modell chapter identifies one reason for the ecological segregation of class and ethnic groups: control, resource allocation, and peer environment of the public schools. Bédarida and Sutcliffe identify another: the behavior of the citizenry in public space, "over which individuals can exercise no direct control."

Beyond being a means of circulation and architectural baseline, the street is considered as a locus of social life, subject to implicit codes of public behavior. The street in nineteenth-century London and Paris was the common ground of a polyglot population of workmen and artisans scurrying to their chores, of mendicants and footpads, of shoppers and boulevardiers. In a few works like the *Life and Death of Great American Cities*, the street is viewed as the protector of norms, the educator of the young, and a place of vitality for residents and tourists. In other works, like *Street Corner Society* or the

Urban Villagers, the street is merely a place to hang out, to see and be seen. But in folklore and most social science literature, the street is seen as a hostile and menacing milieu, in which public codes of behavior are not necessarily shared.

The street is where one encounters strangers, whose trustworthiness is always suspect. Sexton notes that in lower-class Hispanic neighborhoods, parents are constantly worried about their children being beaten, stabbed, or killed. In *Down These Mean Streets, Talley's Corner,* and *Families Against the City,* the variety and excitement appealing to the tourist or boulevardier is positively threatening to stable families. "Street-wise" refers to the possession of unsavory knowledge, while that marvelous co-innovation with the sidewalk, the gutter, connotes to the worst of the worst.

Perhaps there is something ethnic in the exceptions. The most favorable accounts of street life in America describe Italian neighborhoods. The most horrifying accounts describe Black and Puerto Rican neighborhoods. As Jane Jacobs pointed out, Italian-American urban ethos somehow accommodates orderly disorder. Italian street life is the least threatening to strangers, and Italian neighborhoods such as North Beach and Greenwich Village, are sites of farmer's markets and artistic deviants.

These exceptions aside, postwar suburbanization has resulted in a settlement pattern that virtually excludes street life and reduces the fuctions of the street to a mere right of way. The differentiation of the metropolis into zones of homogenous land use reduces the necessity and legitimacy for strangers to enter residential streets. In automobilized metropolitan areas, walking in some residential neighborhoods arouses suspicions of criminal intent.

The desire for a safe, relatively predictable street life lingers and may receive its expression in a new form. The growth of the suburban townhouse development around a lane, the revival of walking malls like Boston's Quincy Market or San Francisco's Cannery, the popularity of touristic Main Streets as in Disney World, or the "gentrification" of older neighborhoods in selected areas may reflect this desire. The success of these revivals apparently depends on privatization of the streets. We expect to hear more about these phenomena from Brian Berry. My suspicion is that neither the potentially violent public street life of older cities nor tedium of suburbs, but bounded unpredictability, is the optimum.

David R. Goldfield:

Within the past half-decade, urban historians have been turning with increasing frequency to the urbanization models and methods devised by urban ecologists and urban geographers. This newer urban history attempts to analyze the relationship between the urban environment and the behavior of urban residents. The form, function, and perception of urban space have become important historiographical digs as scholars probe the process of urbanization. Historians are rather latecomers to this study of space and society, but they liberate geographic and ecological models from their sometimes-ahistorical base.

The blend of urban history on one hand, and urban ecology on the other is evident in the identity of concerns that scholars have assigned to these interconnected disciplines. Theodore Hershberg framed the boundaries of the newer urban history in a recent penetrating essay in the *Journal of Urban History* (November 1978) when he wrote that its content "should consist of the systematic probings of how cities grow, how the urban environment affects behavior and how these two interact." Brian J.L. Berry and John D. Kasarda, in their text, *Contemporary Urban Ecology* (New York, 1977), maintained that the central research issue of urban ecology "is understanding how a population organizes itself in adapting to a constantly changing yet restricting environment"; and, of urban geography, "how and why perceptions of the urban environment produce locational and other choices that reinforce (or change) the spatial processes that are responsible for maintaining (or changing) the urban environment.''

The meshing of history, ecology, and geography and the possibilities as well as the limitations of this union are evident in the two chapters that open this volume. Professor Modell asks whether "an aspect of child rearing [i.e., behavior] helped explain the social geography of the city, i.e., the environment, which in turn accounted for an "urban education boom," i.e., behavior. Considering Professor Modell's collaboration with Theodore Hershberg, this ecological construct is not surprising; neither are the conclusions. Professor Modell takes what we already know about fertility, education, and space, tests it quantitatively, and frames his conclusions in ecological terms. The close relationship between fertility and place of residence, the growing importance of suburbanization as an explicator of urban change, the emergence of familistic wards on the periphery, and the subsequent, though largely unfulfilled, demands for "quality" education are the most interesting results of the study. The findings indicate how behavior and environment can interact to generate urban change.

There are, however, two difficulties with the Hershberg-Modell model concerning the interaction between behavior and environment. First, space can only explain an uncertain portion of behavior. We may expect suburban districts to be increasingly different from central-city areas because the residents are different. Changes in the life cycle, economic opportunity, and status consciousness affect geographic mobility patterns. These factors are bound to vary across the segregated metropolitan areas as particular enviornments present attractions and detractions for each factor. Once residents have placed themselves in certain areas, can we say that the respective environments induced a specific behavioral response—or should we say that these environments were conducive to fulfilling the objectives for which residents chose them in the first place? While environments might alter daily routines and produce changes in lifestyle, do they alter culture?

The post-World War II suburban movement resulted from national demographic patterns, federal legislation, economic conditions, technological advances in the homebuilding industry, and cultural values emphasizing consumption, homeownership, and children after more than a decade of deprivation in all these categories. The new suburbs served as convenient containers into which residents poured their ideals, perceptions, and expectations. As the residents and the environment grew together, the interaction and influence may have become greater, as it did in peripheral Philadelphia between 1880 and 1920. Initially, however, other factors predisposed suburban residents toward emphasizing familistic patterns.

Eric Lampard implied these limits of the urban ecological model when he urged that "the contingencies of events and personalities..., subjective and attitudinal variables," must be taken into account in urban historical research in order to avoid environmental determinism. Planners who believed with nineteenth-century reformer Felix Adler that "squalid houses make squalid people" were rudely disappointed to learn in the 1950s and 1960s that good houses did not necessarily make good people. The culture and economics of poverty considerably outweighed the environmental impact.

Two recent studies, one of architecture, the other of fertility, assign a more passive role to urban space. William H. Pierson's *American Buildings and Their Architects* (Garden City, NY, 1978) suggests that technology formed nineteenth-century New England industrial architecture, and that English literature, painting, and landscaping inspired the Gothic revival style in America. Technology and culture—variables that transcended the urban context, and urban space for that matter—nurtured distinctive urban architectural forms. These forms appeared in specific urban spaces, of course, but again, space acted as receptacle, not as activator.

Peter Lindert's *Fertility and Scarcity in America* (Princeton, 1978) posits that fertility rates during the past century depended on the childhood experiences and economic calculations of the prospective parents. Perhaps Professor Modell's more prolific suburbanites were reacting to *their* own childhood familial experiences and to their recently gained affluence in reaching childbearing and child rearing decision.

The second problem with the ecological model is that behavior and environment are not that separable. Urban space—its disposition and evolution—is itself entwined with behavioral responses. Real-estate owners and developers, banks, transit companies, and public authorities have significant roles in molding the spatial container. Sam Bass Warner, Jr.'s analysis of land-use evolution in peripheral Boston emphasized the role of the private entrepreneurial decision makers in shaping the suburbs. Thomas J. Anton, a University of Michigan political scientist, suggested that studying the actors in the drama of urban land-use development—their motives, philosophical backgrounds, and constraints (economic, political, and cultural)—would produce some important insights into the urban process. Perhaps it might be more appropriate to speak of how the behavior of spatial development affects the behavior of settlement—the choices made by those suburban residents and businesses discussed by Professor Modell.

The ecological model demonstrates similar promise and limitation in the Bédarida-Sutcliffe essay on the street in the structure and life of the modern city. The relationships between streets, buildings, and the people inhabiting those structures have not been explored systematically. As Bédarida and Sutcliffe suggest, these relationships reveal important aspects of the urban process. The tripartite connection between structure, street, and individual shifted over the centuries, until finally, in the twentieth century, internal space had been so improved and external space so degraded that the socially interactive functions of streets were reduced considerably. In the modern American suburb, garage doors are the faces residents show to the streets. The familistic settings emphasize the home and the backyard as the environment of primary interaction. The street's social and commercial functions have shifted to the shopping center. The suburban street is simply a vehicular thoroughfare. Incidentally, the shopping center may not be so much an evolution of a new form as a modern rendition of the medieval market place—a device that evolved, in part, from the inefficiency of the streets as commercial thoroughfares.

The street scenario points to other and larger factors in determining behavioral patterns than the streets or structures themselves. The abandonment of the street in the twentieth century resulted from broad cultural (emphasis on family and home), economic (affluence to make the privatization of space feasible), and technological (transportation and communication) events occurring in regional and national contexts. Richard Sennett chronicles the withdrawal of the American urban man—his avoidance of conflict, diversity, and change—in *Uses of Disorder*. The American urban man built his environment to reflect these new social preferences. The absence of sidewalks was no longer a sign of primitiveness; it was a matter of sophisticated choice.

Themes beyond the physical environment were evident considerably earlier in the evolution of street form and function. As in suburban Philadelphia, political decisions were also superimposed on the empty container—the street. Royal choices in favor of wide thoroughfares in Paris expanded options as to what could be done in these new spaces, and residents, strollers, and merchants filled the container with their choices accordingly. In London, the central political direction of street and house building evident in Paris was generally absent. The home and the garden were more attractive choices than the streets, which, as Bédarida and Sutcliffe noted, "offered little excitement or distraction."

The limits of ecological perspective are evident in these cross-national comparisons. Aside from the influence of urban space, it is possible that French culture is more oriented toward and attuned to street life than English culture. In southern European cities, street life is endemic whether on broad boulevards or on less pretentious space. The distinction between public and private space is less sharp. The home is more public in southern European cultures, as is the street. Are housing conditions, more primitive technology, and the absence of interior space responsible for this street-life proclivity, or are more basic cultural values rooted in familial traditions, for example, a better explanation of Neopolitan street life? Are the Neopolitans "forced out into the street," or is it a preferred choice?

In Stockholm, for another example, there is very little street life. Physically, the city is closer to Paris than to London—broad broulevards; few, narrow, and quaint passageways (outside the Old Town); and a housing stock as bulky and including units as small as Parisian counterparts. Weather certainly plays a role here, but there is very little neighboring as well. The elimination of courtyards in apartment construction during the 1930s received architectural praise and little popular criticism. Stockholm planners, attempting to create social spaces in peripheral areas after World War II, admitted failure by the mid-1950s.

Perhaps the most extreme example of the influence of culture over environment is a street in Grenoble, France, discussed in Donald Appleyard's *Urban Conservation in Europe and America* (Rome, 1975). Algerians, French, and Italians shared the housing, but not the street. The impression is that the quarter is entirely Algerian because they predominate on the street. The Italians confine their interaction to their flats, where neighboring is common, and occasionally to the street. Finally, the French, primarily an elderly female population, are rarely visible in the street and lead generally solitary existences in their residences. Thus, on the same street with virtually the same housing stock there were three different responses generated by cultural and life-cycle distinctions.

Before we apply ecological models to cross-national studies, we must analyze cultural patterns of social interaction and familial traditions. Culture is not oblivious to space, of course, but based on this brief survey it may be a more significant determinant of behavior. The fact that Londoners "muddle through without taking up definite positions for or against the street," while Parisians have reacted strongly against Corbusian intrusions, may have more to do with culture than with space.

The ecological framework for urban historical analysis, by asking new questions, has produced and will produce significant advances in our knowledge of the urban process and, by implication, of how cities function today and how they can function better tomorrow. The foregoing discussion suggests that the ecological model of behavior and environment involves a partial explanation of the historical process of urban change.

Consider the case of Francie Nolan, a young Irish girl from Brooklyn. How to fit Francie (or Brooklyn for that matter) into the ecological framework? The young heroine of Betty Smith's novel, *A Tree Grows in Brooklyn*, grew up amid the cracked asphalt, broken panes, and fallen plaster of a relic of a neighborhood. Her quest for knowledge, her reading on the rusty fire escape, her college education and subsequent success were obviously not environmentally inspired. What accounted for Francie's triumph, then? She was, according to Betty Smith,

> the books she read in the library. She was of the flower in the brown bowl. Part of her life was made from the tree growing rankly in the yard. She was the bitter quarrels she had with her brother whom she loved dearly. She was Katie's secret, despairing weeping. She was the shame of her father staggering home drunk. She was all of these things and of something more that did not come from the Rommelys nor the Nolans, the reading, the observing, the living from day to day. It was something that had been born into her and her only—the something different from anyone else in the two families. It was what God or whatever is His equivalent puts into each soul that is given life—the one different thing such as that which makes no two fingerprints on the face of the earth alike.

All of these things were part of Francie's "environment," to be sure. But ecologists have a much more narrow definition of environment: physical space. There was the tree, but the ecological model cannot see the tree for the forest of space. Urban historians, by the nature of their subject, though, cannot focus on the individual urban resident. Urban space is a more convenient, manageable, and quantifiable unit of research. It is important as an organizing framework for research on urban process. What is at question here is its relative influence on urban change in relation to other factors within and without the urban environment. Locational decisions are the essence of urban change. The reasons they occur, however, may be due less to the location than to the economic, demographic, political, cultural, and technological constraints or conductors present in urban, regional, and national societies.

II

CLASS TENSION AND THE MECHANISMS OF SOCIAL CONTROL: THE HOUSING EXPERIENCE

The two essays and commentary in this section consider the use of housing as a mechanism of social control. Kenneth T. Jackson studies the impact of U.S. government housing policy between 1918 and 1968 in an attempt to determine whether the results of such policy were "foreseen by a government anxious to use its power and resources for the social control of ethnic and racial minorities." Lutz Niethammer investigates the debate over nineteenth-century European housing reform in a similar manner, but attempts a broader analysis of the nature of industrial capitalism.

After a brief discussion of government and housing prior to 1933, Jackson emphasizes the Home Owners' Loan Corporation, the Federal Housing Administration, and the U.S. public housing program. The HOLC created a formal and uniform system of appraisal that led to the practice of "redlining"—the undervaluing of neighborhoods that were dense, mixed, or aging areas that may have been occupied by ethnic and racial groups labeled "undesirable." As Jackson indicates, HOLC assumptions about urban neighborhoods were based on both an ecological conception of change and a socioeconomic one. The New Deal appraisers accepted the inevitability of change in the American city and the natural tendency of neighborhood decline because of structural aging and obsolescence, along with the moving in of lower-income and "less desirable" ethnic or racial groups. For them, physical deterioration was both the cause and the effect of population change. Areas listed as low-grade were redlined because of age and because they were "within such a low price or rent range as to attract an undesirable element."

The FHA, which insured mortgage loans of private lenders, revolutionized the home finance industry. Especially after World War II, through long-term, low-interest, fixed-rate mortgages, the FHA encouraged suburbanization, hastened inner-city neighborhood decay, and sapped cities of their white middle-class population. In this respect, government policy hastened the phenomenon that became known as "white flight." Unlike the HOLC, which apparently disregarded some of its own guidelines and extended aid to low-rated neighborhoods, the FHA followed HOLC appraisal practices and clearly favored suburban areas over industrial cities. Hence, while other factors such as family size, attitudes toward child rearing, or stages in the life cycle cannot be neglected in considering movement to the suburbs, neither can government policy.

While government mortgage policy intensified flight from the city, New Deal public housing programs not only marked a distinct reversal of traditional federal policy, but initiated developments that would have a severe impact on America's older industrial cities. For a variety of reasons explained by Jackson, public housing worked to institutionalize urban ghettos and further to socially divide city from suburb in the United States. In all of this, one must consider, as commentator Peter Marcuse suggests, whether the federal government led or followed broader private forces in the process of suburbanization. Without government activity such as that described by Jackson, would the urban-suburban configuration of our society have been any different? One must also ask, as does Christine Rosen, whether the policy makers and bureaucrats who created the programs did so with the intention of using them as instruments of social control or as a means of minimizing investment risk. Moreover, if the intent was social control, what *kind* of social control did such programs promote? The question, itself, raises the issue of policies and reforms that intend to stabilize and absorb, perhaps co-opt, a group, and those that intend to isolate and contain. Hence, the historian must consider the difficult twin issues of intent or motivation of the policy maker or reformer, on one hand, and the type of social control that is imposed, on the other.

For Lutz Niehammer, the nineteenth-century European debate over housing reform is viewed as an attempt to moderate the class struggle; it is, for him, a new paradigm of social control. He discusses the evolution of this paradigm in three ways: as a theoretical overview of the basic cultural dialectic in urban society during the rise of industrial capitalism; through an analysis of the writings of Edwin Chadwick, James Hobrecht, and Frederick LePlay, which considers their ideas concerning health, space, and family; and by seeing it as a method that attempted to break the economic deadlock preventing any real housing reform in the nineteenth century. As bourgeois homes grew increasingly differentiated in terms of private space, bourgeois horror, according to Niethammer, increased about the "savage" fluidity of working-class "open housing." As a consequence, the view developed that the poor working class was savage and immoral, and the middle class felt increasingly threatened as cities grew with the migration of large numbers of impoverished Europeans to them. Even the most liberal bourgeois reformers came to accept the need for authoritarian control of the urban poor. Housing, like public health reform, and the campaign to maintain the family as a cohesive unit emerged as vehicles to stabilize society and ensure the progress of industrial capitalism.

One conference participant, Michael McCarthy, suggested that the motivation of a reformer such as Chadwick had been misread. He claimed that Chadwick had to employ a moralistic approach in order to have his ideas accepted by the Parliament and the bourgeoisie. Hence, humanitarianism rather than the maintenance of stability served as his and other reformers' motivation. Claiming that such moralism was ideolgoical rather than pragmatic, Niethammer countered that the reformers' arguments represented a way of thinking rather than a mode of salesmanship. This disagreement, nevertheless, highlights the difficulty in determining motivation when considering reform efforts and, thereby, touches on a major historiographic debate among historians. Moreover, Rosen suggests that both Jackson and Niethammer approach the issue of housing and social control "from the top down"—that is, they treat housing policy as being manipulated by elites who impose such policy on those who have little power to resist. Her assertion that one must consider the reaction and views of those being *acted upon* adds an important dimension to the argument. The study of working-class culture within the urban environment offers not only a challenge but also an especially fruitful area of consideration.

Thus, as one ponders current trends in the development of America's modern industrial cities, scholars, as well as policy makers, must concern themselves not only with the activities of housing reformers and government bureaucrats, but with the urban poor *and* the middle class. An economy that makes the inner city more appealing to the children of those who left it for the suburbs, may work to change the social geography of urban and suburban America. Federal policy that may encourage the process of "gentrification" also may work to displace the urban poor, who, in Jackson's words, "will pay the price." As he notes, "the South African experience indicates social control does not require that ghettos invariably be located at the center." Moreover, an economy that makes homeownership more difficult for everyone may, if such ownership actually does encourage social stability, have long-term implications for our society. Futher discussion of such implications will be found in Part III.

4

THE SPATIAL DIMENSIONS
OF SOCIAL CONTROL

Race, Ethnicity, and Government Housing Policy in the United States, 1918-1968

KENNETH T. JACKSON

If a healthy race is to be reared, it can be reared only in healthy homes; if infant mortality is to be reduced and tuberculosis to be stamped out, the first essential is the improvement of housing conditions; if drink and crime are to be successfully combated, decent sanitary houses must be provided. If "unrest" is to be converted into contentment, the provision of good houses may prove one of the most potent agents in that conversion.
—King George V, April 11, 1919

A nation of homeowners, of people who own a real share in their own land, is unconquerable.
—President Franklin D. Roosevelt

Suburbanization is a process that began in the United States in the first quarter of the nineteenth century and has continued at least through the 1970s. The appeal of low-density living over time and across regional, class, and ethnic lines has led some observers to regard it as natural and inevitable, a trend "that no amount of government interference can reverse."[1] As a senior Federal Housing Administration (FHA) official told the 1939 convention of the American Institute of Planners, "Decentralization is taking place. It is not a policy, it is a reality—and it is as impossible for us to change this trend as it is to change the desire of birds to migrate to a more suitable location."[2]

Despite such protestations, there are many ways in which government largesse can affect where people live. For example, the existing tax code encourages businesses to abandon old structures too soon. When this is compounded by permitting greater tax benefits for new construction than for the improvement of existing structures, the government subsidizes an acceleration in the rate at which economic activity is dispersed to new locations.[3] Similarly, it has recently been demonstrated that defense spending has been encouraging the growth of the Sunbelt. In New York City alone, the Regional Plan Association estimated that Washington collected $6 billion more than it spent in 1975 in the metropolitan area.[4]

On the urban-suburban level, the potential for federal influence is also enormous. For example, the Federal Highway Act of 1916 and the Interstate Highway Act of 1956 moved the government toward a transportation policy emphasizing and benefiting the road, the truck, and the private motor car.[5] In conjunction with cheap fuel and mass-produced automobiles, the urban expressways led to lower marginal transport costs and greatly stimulated decentralization. Even the reimbursement formulas for water-line and sewer construction have had an impact on the spatial patterns of metropolitan areas.[6]

The purpose of this chapter is to examine the impact of federal housing policies on older, industrial cities and on newer, suburban areas. More specificially, I seek to determine whether the results of those policies were foreseen by a government anxious to use its power and resources for the social control of ethnic and racial minorities. Has the American government been as benevolent—or at least as neutral—as its defenders have claimed?[7]

GOVERNMENT AND HOUSING BEFORE 1933

Although housing involves the largest capital costs of any human necessity, for the first three centuries of urban settlement in North America the provision of shelter was not regarded as a responsibility of government—whether that body was a colonial assembly or a state legislature, a town meeting or a city council, a Parliament in London or a Congress in Washington. Local governments occasionally outlawed wooden dwellings and thatched roofs in city centers as early as the seventeenth century,[8] and New York City passed restrictive housing laws as early as 1867, but the selection, construction, and purchase of a place to live was everywhere regarded as an essentially individual problem. Prior to the 1930s, federal involvement was limited to a survey of slum conditions in large cities in 1892, the creation of the Federal Land Bank System in 1916, and the construction of arms workers' housing during World War I.[9]

This last and potentially most important shift came in June 1918, when Congress appropriated $100 million to form the United States Housing Corporation. The purpose was to provide residences for heads of households migrating to industrial areas in order to produce munitions and ships for the European conflict. But because this war emergency effort began only five months before the Armistice, it resulted in only a few developments—in Bridgeport, Connecticut; Camden, New Jersey; Portsmouth, New Hampshire; Wilmington, Delaware; Kohler, Wisconsin; and Chester, Pennsylvania. At the cessation of hostilities, some suggested that the national government maintain the structures for the benefit of low-income workers, but the constitutionality of such a venture was considered questionable, and private developers eventually bought the units.[10]

At the same time that Washington remained absent from the housing field, European governments charted a new course.[11] Both Great Britain and Germany built more than one million publicly assisted dwelling units between 1920 and 1930. In The

Netherlands, the government rehoused one-fifth of the total population in the same fashion, while in the Soviet Union the transition to public responsibility was almost total. As the American housing reformer Edith Elmer Wood noted sadly in 1931, "Nearly all other European countries have developed some form of housing loan at low interest rates and some form of municipal housing or a thinly disguised substitute for it." Great Britain, according to her estimate, was a half century ahead of the United States in the field of shelter.[12]

With the advent of the Great Depression in 1929, however, the American attitude toward housing began to change in a fundamental way. The sharp economic downturn inflicted crippling blows on both the housing industry and the home-owner. Between 1928 and 1933, the construction of residential property fell by 95 percent, and expenditures on home repairs fell by 90 percent. In 1930, about 150,000 nonfarm home-owners lost their property through foreclosure; in 1931, this increased to nearly 200,000; in 1932, to 250,000. According to federal estimates in 1933, fully half of all home mortgages in the United States were technically in default, and foreclosures reached the astronomical rate of more than a thousand per day. Housing prices predictably declined, virtually wiping out vast holdings in second and third mortgages as values fell below even the primary claim. Moreover, the victims were often middle-class families who were experiencing impoverishment for the first time.[13]

Theorizing that the predicament of the real-estate and construction industries was acting as a drag on the rest of the economy, and believing that homeownership was "both the foundation of a sound economic and social system and a guarantee that our country will continue to develop rationally as changing conditions demand," Herbert Hoover convened the President's Conference on Home Building and Home Ownership in 1931.[14] In an address at the opening meeting, President Hoover gave expression to the national preference for the private house.

I am confident that the sentiment for home ownership is so embedded in the American heart that millions of people who dwell in tenements, apartments, and rented rooms...have the aspiration for wider opportunity in ownership of their own homes.[15]

The conference made four recommendations that pointed to a new direction in federal housing policy: (1) the creation of long-term, amortized mortgages;[16] (2) the encouragement of low interest rates; (3) the institution of government aid to private efforts to house low-income families; and (4) the reduction of home construction costs. And the conference closed with a warning:

> This committee is firmly of the opinion that private initiative taken by private capital is essential, at the present time, for the successful planning and operation of large-scale projects. Still, if we do not accept this challenge, the alternative may have to be government housing.[17]

The Hoover administration tried to encourage homeownership in two ways. It established the Federal Home Loan Bank Board in 1932 to serve as a credit reserve for mortgage lenders and thus to increase the supply of capital in the housing market. But it was not designed to give help in cases of emergency distress and was able to give aid only where the risk was slight. Not surprisingly, it was ineffective, and conditions became appreciably worse. A second measure, the Emergency Relief and Construction Act of 1932, also proved inconsequential. It empowered the Reconstruction Finance Commission to

> make loans to corporation formed wholly for the purpose of providing housing for families of low income, or for the reconstruction of slum areas, which are regulated by state or municipal law as to rents, capital structure, rate of return, and areas and methods of operation, to aid in financing such projects undertaken by such corporation which are self-liquidating in character.[18]

Unfortunately, the act required the states to exempt such limited-dividend corporations from all taxes, and at the time only New York had such authority. As a result, Knickerbocker

Village in New York City was the only project initiated under the legislation.

It remained for Franklin D. Roosevelt and his Democratic majority to develop new initiatives in housing. One of the freshest efforts of the New Deal was the Greenbelt Town Program. Inspired by Rexford G. Tugwell and administered through his Resettlement Administration, the purpose of the program was explicitly to foster deconcentration. Tugwell intended to construct ideal "greenbelt" communities based on the planning theories of England's Ebenezer Howard and then to sell the projects to private enterprise. As Tugwell explained it, "My idea was to go just outside centers of population, pick up cheap land, build a whole community, and entice people into them. Then go back into the cities and tear down whole slums and make parts of them."

The Greenbelt Town Program came under vigorous conservative attack, however. A proposed New Jersey community never even made it off the drawing board, and the three garden communities that were completed—Greenbelt in Maryland, Greenhills in Ohio, and Greendale in Wisconsin—were hurt by excessive construction costs and never served as models for future metropolitan development.[19]

Three other innovations of the New Deal—the Home Owners' Loan Corporation, the Federal Housing Administration, and the public housing program—were to have a more lasting impact on the nation and its people.

THE HOME OWNERS' LOAN CORPORATION

On April 13, 1933, President Roosevelt sent both the House and the Senate a message urging passage of a law that would (1) protect the small homeowner from foreclosure, (2) relieve him of part of the burden of excessive interest and principal payments incurred during a period of higher values and higher earning power,[20] and (3) declare that it was national policy to protect homeownership. The resulting Home Owners' Loan Act was one of the first measures passed by the new 73rd Congress and was signed by the President on June 13, 1933.[21]

The HOLC was designed to refinance mortgages in danger of default or foreclosure, and even to make loans to permit owners to recover homes lost through forced sale. Between July, 1933, and June, 1935, alone, the HOLC supplied more than $3 billion for over one million mortgages, or loans for one-tenth of all owner-occupied, nonfarm residences in the United States.[22] Although applications varied widely by state (in Mississippi, 99 percent of the eligible owner-occupants applied for loans, while in Maine only 18 percent did so), Professor C. Lowell Harriss estimated that nationally, about 40 percent of eligible Americans sought HOLC assistance.

Aside from the large number of mortgages it helped to refinance on a long-term, low-interest basis, the HOLC is important to housing history because of its systemization of appraisal methods. Because it was dealing with problem mortgages—in some states over 40 percent of all HOLC loans were foreclosed even after refinancing—the HOLC had to make predictions and assumptions regarding the useful or productive life of housing it financed. Unlike refrigerators or shoes, dwellings were expected to be durable; how durable was the purpose of the investigation.

With care and extraordinary attention to detail, HOLC appraisers divided cities into neighborhoods and developed elaborate questionnaires relating to the occupation, income, and ethnicity of the inhabitants and the age, type construction, price range, sales demand, and general state of repair of the housing stock. In evaluating such efforts, the distinguished economist C. Lowell Harriss has credited the HOLC training and evaluation procedures "with having helped raise the general level of American real estate appraisal methods."[23] A less favorable judgment would be that the Home Owners' Loan Corporation initiated the practice of "redlining."[24]

This occurred because HOLC devised a rating system that undervalued neighborhoods that were dense, mixed, or aging. Four categories of quality—imaginatively entitled, First, Second, Third, and Fourth, with corresponding code letters of A, B, C, and D and colors of green, blue, yellow, and red— were established. The First or A grade (green) areas were described as new, homogeneous, and "in demand as residential locations in good times and bad." Homogeneous meant "American business and professional men." Jewish neighborhoods or even those with an "infiltration of Jews" could not be considered "best."[25]

The Second security grade (blue) went to "still desirable" areas that had "reached their peak," but were expected to remain stable for many years. The Third grade (yellow) or C neighborhoods were "definitely declining" because of age, obsolescence, or change of style. "Having seen their better days," such yellow-colored sections were "within such a low price or rent range as to attract an undesirable element." Finally, the Fourth grade (red) or "hazardous" areas were those "in which the things taking place in C areas have already happened." Black neighborhoods were invariably rated D, as were any areas characterized by poor maintenance, poverty, or vandalism.[26]

The Home Owners' Loan Corporation did not develop the idea of considering race and ethnicity in real-estate appraising.[27] As Calvin Bradford has demonstrated, models developed at the University of Chicago in the 1920s and early 1930s by Homer Hoyt and Robert Park became the dominant explanation of neighborhood change.[28] They suggested that different groups of people "infiltrated" or "invaded" territory held by others through a process of competition. These interpretations were then picked up by prominent appraising texts, such as Frederick Babcock's *The Valuation of Real Estate* (1932) and *McMichael's Appraising Manual* (1931). Both of them advised appraisers to pay particular attention to "undesirable" or "least desirable" elements and suggested that the influx of certain ethnic groups was likely to precipitate price declines.

But the Home Owners' Loan Corporation did apply these notions of ethnic and racial worth to real-estate appraising on an unprecedented scale. With the assistance of local realtors and banks, the HOLC assigned one of its four ratings to every block in every city. The resulting information was then tran-

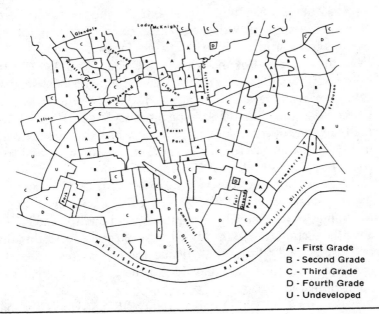

Figure 4.1: St. Louis Area Residential Security Map, 1937

SOURCE: Record Group 31, National Archives.

slated into the appropriate color and duly recorded on secret "residential security maps" in local HOLC offices.[29] The following information is based on the maps and neighborhood evaluations pertaining to the St. Louis and Newark metropolitan areas.[30]

The residential security maps for the St. Louis area, as Figure 4.1 indicates, gave the highest ratings to the newer, affluent neighborhoods that were strung out along curvilinear streets well away from the problems of the city. At the top of the scale was Ladue, a large undeveloped section of high, rolling land and heavily wooded estates. In 1940, federal appraisers noted approvingly that the area was "highly restricted"[31] and occupied by "capitalists and other wealthy families." Reportedly, Ladue was not home to a single foreigner or Negro; its rating was A. Other affluent suburbs like Clayton, University City, and Webster Groves were also marked with green and blue on the 1937 map, indicating that they, too, were characterized by attractive homes on well-maintained plots, and that the appraisers felt confident about mortgages insured there.

At the other end of the scale in St. Louis County were the rare Fourth grade areas. A few such neighborhoods were occupied by white laborers, such as "Ridgeview" in Kirkwood, where the garage-like shacks typically cost less than $1500. But the D regions of the county were usually black. One such place in 1937 was Lincoln Terrace, a small enclave of four- and five-room bungalows built in 1927. Originally intended for middle-class white families, the venture was unsuccessful, and the district developed into a black neighborhood. But even though the homes were relatively new and of good quality, the HOLC gave the section (D-12) the lowest possible grade, asserting that the houses had "little or no value today, having suffered a tremendous decline in values due to the colored element now controlling the district."[32]

In contrast to St. Louis County, St. Louis City had proportionately many more Third and Fourth grade neighborhoods.[33] As Figure 4.1 indicates, virtually all the residential sections along the Mississippi River or adjacent to the central business district (CBD) received the lowest two ratings.[34] The evaluation of a white, working-class neighborhood near Fairgrounds Park was typical. According to the area description, "Lots are small, houses are only slightly set back from the sidewalks, and there is a general appearance of congestion." Although an urbane individual might have found this collection of cottages and abundant shade trees rather charming, the HOLC thought otherwise: "Age of properties, general mixture of type, proximity to industrial section on northeast and much less desirable areas to the south make this a good Fourth grade area."

As was the case in the county and in other cities, any Afro-American presence was a source of substantial concern to the federal appraisers. In a confidential and generally pessimistic 1941 survey of the economic and real-estate prospects of the St.

Louis metropolitan area as a whole, the Federal Home Loan Bank Board (the parent agency of HOLC) repeatedly commented on the "rapidly increasing Negro population" and the resulting "problem in the maintenance of real-estate values." The officials evinced a keen interest in the movement of black families and included maps of the density of Negro settlement with every analysis. Nor surprisingly, black neighborhoods were invariably rated D, or hazardous.[35]

Like St. Louis, Newark has long exemplified the most extreme features of the urban crisis. In that troubled city, federal appraisers took note in the 1930s of the high tax rate, the heavy relief load, the per-capita bonded debt, and the "strong tendency for years for people of larger incomes to move their homes outside the city."[36] As Figure 4.2 indicates, the 1939 Newark area residential security map did not designate a single neighborhood in that city of more than 400,000 as worthy of an A rating. "High-class Jewish" sections like Weequahic and Clinton Hill, as well as gentile areas like Vailsburg and Forest Hill all received B, the Second grade.[37] Typical Newark neighborhood were rated even lower. The well-maintained and attractive working-class sections of Roseville, Woodside, and East Vailsburg were given Third grade or C ratings; the remainder of the city, including places like all-white but immigrant Ironbound and all the black neighborhoods, were largely written off as hazardous.

Immediately adjacent to Newark is New Jersey's Hudson County, which is among the half-dozen most densely settled and ethnically diverse political jurisdictions in the United States. Predictably, HOLC appraisers had decided by 1940 that Hudson County was a lost cause. In the communities of Bayonne, Hoboken, Secaucus, Kearny, Union City, Weehawken, Harison, and Jersey City, taken together, they designated only two very small B areas and no A sections.

The Home Owners' Loan Corporation insisted that "there is no implication that good mortgages do not exist or cannot be made in Third and Fourth grade areas."[38] And, as Table 4.1 indicates, the HOLC did in fact make the majority of its

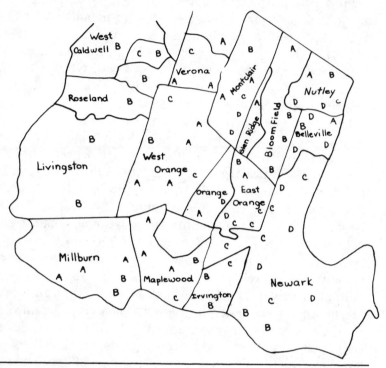

Figure 4.2: Residential Security Map of Essex County, New Jersey, as Prepared by the Federal Home Loan Bank Board as of June 1, 1939

obligations in "definitely declining" or "hazardous" neighborhoods. This seeming liberality was actually good business, because the residents of poorer sections generally maintained a better pay-back record than did their more affluent cousins. As the Federal Home Loan Bank Board explained,

The rate of foreclosure per 1000 non-farm dwellings during 1939 was greater in St. Louis County than in St. Louis City by about 2½ to 1. A partial explanation or causation of this situation is the fact that County properties consist of a greater proportion of units in the higher-priced brackets.[39]

TABLE 4.1   Distribution of HOLC Loans in Essex County (Newark),
             New Jersey, and Shelby County (Memphis), Tennessee,
             According to Neighborhood Classifications, 1935—1936

Classification	Essex County Number	Essex County Percentage	Shelby County Number	Shelby County Percentage
A: Best	685	10.2	129	4.7
B: Still desirable	1,975	29.3	752	27.6
C: Definitely declining	2,156	32.0	1,003	36.8
D: Hazardous	1,917	28.5	843	30.9

SOURCE: Compilations made from HOLC and FHA reports in Record Group 195, National
Archives.

The damage caused by the HOLC came not through its own
actions, but through the influence of its appraisal system on the
financial decisions of other institutions. During the late 1930s,
the Federal Home Loan Bank Board circulated questionnaires
to banks asking about their mortgage practices. Those returned
by savings and loan associations and banks in Essex County
(Newark), New Jersey, indicated a clear relationship between
public and private "redlining" practices. One specific question
asked, "What are the most desirable lending areas?" The
answers were often "A and B" or "blue" or "FHA only."[40]
Similarly, to the inquiry, "Are there any areas in which loans
will not be made?" the responses included "red and most
yellow," "C and D," "Newark," "not in red," and "D
areas." Obviously, private banking institutions were privy to
and influenced by the government's residential security maps.[41]
Even more significantly, HOLC appraisal methods, and
perhaps the maps themselves, were adopted by the Federal
Housing Administration.

THE FEDERAL HOUSING ADMINISTRATION

Direct, large-scale, Washington intervention in the American
housing market dates from June 27, 1934, with the passage of
the National Housing Act "to encourage improvement in
housing standards and conditions, to facilitate sound home

financing of reasonable terms, and to exert a stabilizing influence on the mortgage market.'' Congressmen were especially
hopeful that it would alleviate unemployment in the construction industry. As the Federal Emergency Relief Administrator testified on May 18, 1934, before the Banking and
Currency Committee,

> The building trades in American represent by all odds the largest
> single unit of our unemployment. Probably more than one-third
> of all the unemployed are identified, directly and indirectly, with
> the building trades.... Now, a purpose of this bill, a fun-
> damental purpose of this bill, is an effort to get these people
> back to work.

The FHA effort was later supplemented by the Servicemen's
Readjustment Act of 1944 (more familiarly known as the GI
Bill), which created a Veterans Administration (VA) program
to help the fifteen million GIs of World War II return to
civilian life with a home of their own.[42]

Both the FHA and the VA housing programs have had a
remarkable record of accomplishment. Essentially, they insure
long-term mortgage loans made by private lenders for home
construction and sale. To this end, they collect premiums, set
up reserves for losses, and in the event of a default on a
mortgage, indemnify the lender. Neither the FHA nor the VA
extend credit or build houses. What they have done is to
revolutionize the home finance industry in the following
ways.[43]

(1) Before FHA and VA began operation, first mortgages
typically were limited to one-half or two-thirds of the appraised
value of the property. During the 1920s, for example, savings
and loan associations held one-half of America's outstanding
mortgage debt. Those mortgages averaged 58 percent of the
estimated property value.[44] Thus, prospective homebuyers
needed a down payment of at least 30 percent to close a deal. By
contrast, the fraction of the collateral that the lender was able
to lend for an FHA secured loan was about 93 percent, and for
a VA loan about 98 percent. Thus, large down payments were
unnecessary.

(2) Prior to the 1930s, the typical length of a mortgage was between five and ten years, and the loan itself was not fully amortized.[45] Thus, the homeowner was periodically at the mercy of the arbitrary and unpredictable forces in the money market. When money was easy, renewal every five or seven years was no problem. But if a mortgage expired at a time when money was tight, it might be impossible for the homeowner to secure a renewal, and foreclosure would ensue. Under the FHA and VA programs, continuing a trend begun by the Home Owners' Loan Corporation, the loans were fully amortized, and the repayment period was extended to twenty-five or thirty years. The effect was to reduce both the average monthly payment and the national rate of mortgage foreclosure. The latter declined from 250,000 nonfarm units in 1932 to only 18,000 in 1951.

(3) In the 1920s, the interest rate for first mortgages averaged between 6 and 8 percent. If a second mortgage were necessary, as it usually was for families of moderate incomes, the purchaser could obtain one by paying a discount to the lender, a higher interest rate on the loan, and perhaps a commission to a broker. Together, these charges added about 15 percent to the purchase price. Under the FHA and VA programs, by contrast, there was very little risk to the banker if a loan turned sour. Reflecting this government guarantee, interest rates fell by two or three percentage points.[46]

Together, these three changes substantially increased the number of American families who could reasonably expect to purchase homes. By the end of 1972, FHA had helped nearly eleven million families to own houses and another twenty-two million families to improve their properties. It had also insured 1.8 million dwellings in multiunit projects. And in those same years between 1934 and 1972, the percentage of American families living in owner-occupied dwellings rose from 44 percent to 63 percent (see Table 4.2).[47]

Quite simply, it often became cheaper to buy than to rent. Long Island builder Martin Winter recently recalled that in the early 1950s, families living in the Kew Gardens section of Queens were paying almost $100 per month for small, two-

TABLE 4.2   New Housing Starts in the United States, 1935-1968 (in thousands)

Year	Total Starts	FHA Starts	VA Starts	Public Housing
1935	216	14	0	5
1936	304	49	0	15
1937	332	60	0	4
1938	399	119	0	7
1939	458	158	0	57
1940	530	180	0	73
1941	619	220	0	87
1942	301	166	0	55
1943	184	146	0	7
1944	139	93	NA	3
1945	325	41	9	1
1946	1015	69	92	8
1947	1265	229	160	3
1948	1344	294	71	18
1949	1430	364	91	36
1950	1408	487	191	44
1951	1420	264	149	71
1952	1446	280	141	59
1953	1402	252	157	36
1954	1532	276	307	19
1955	1627	277	393	20
1956	1325	192	271	24
1957	1175	168	128	49
1958	1314	295	102	68
1959	1495	332	109	37
1960	1230	261	75	44
1961	1285	244	83	52
1962	1439	259	78	30
1963	1582	221	71	32
1964	1502	205	59	32
1965	1451	196	49	37
1966	1142	158	37	31
1967	1268	180	52	30
1968	1484	220	56	38

SOURCE: U.S. Bureau of the Census, *Housing Construction Statistics, 1889-1964* (Washington, DC: Government Printing Office, 1966), Table A-2; and U.S. Department of Housing and Urban Development, *HUD Trends: Annual Summary* (Washington, DC: HUD, 1970).

bedroom apartments. For less money they could, and often did, move to the new Levittown-type developments springing up along the highways from the city.[48] Even the working classes could aspire to homeownership. As one person who left New York for suburban Dumont, New Jersey, remembered, "We had been paying $50 a month rent, and here we come up and live for $29 a month. That paid everything—taxes, principal, insurance on your mortgage, and interest."[49]

Unfortunately, the corollary to this achievement was the fact that FHA and VA programs hastened the decay of inner-city neighborhoods by stripping them of much of the middle-class constituency.[50] This occurred for two reasons. First, although the legislation nowhere mentioned an antiurban bias, it favored the construction of single-family and discouraged construction of multifamily projects through unpopular terms.[51] Similarly, loans for the repair of existing structures were small and for short duration, which meant that families of modest circumstances could more easily finance the purchase of a new home than the modernization of an old one.[52] Finally, the only part of the 1934 act relating to low-income families was the embryonic authorization for mortgage insurance with respect to rental housing in regulated projects of public bodies or limited dividend corporations. Almost nothing was insured until 1938, and even thereafter, the total insurance for rental housing exceeded $1 billion only once between 1934 and 1962.[53]

The second and more important variety of suburban, middle-class favoritism had to do with the "unbiased professional estimate" that was a prerequisite of any loan guarantee.[54] This mandatory appraisal included a rating of the property itself, a rating of the mortgagor or borrower, and a rating of the neighborhood. The lower was the valuation placed on properties, the less was government risk and the less generous was the aid to the potential buyers (and sellers).[55] The purpose of the neighborhood evaluation was "to determine the degree of mortgage risk introduced in a mortgage insurance transaction because of the location of the property at a specific site."[56] And unlike the Home Owners' Loan Corporation, which used an essentially similar procedure, the Federal Housing Administration allowed personal and agency bias in favor of all-white subdivision in the suburbs to affect the kinds of loans it guaranteed—or, equally important, refused to guarantee. In this way the bureaucracy influenced the character of housing at least as much as the 1934 enabling legislation did.

The Federal Housing Administration was quite precise in teaching its underwriters how to measure the quality of a residential area. Eight criteria were established (the numbers in parentheses reflect the percentage weight given to each):[57]

(a) relative economic stability (40)
(b) protection from adverse influences (20)
(c) freedom from special hazards (5)
(d) adequacy of civic, social, and commercial centers (5)
(e) adequacy of transportation (10)
(f) sufficiency of utilities and conveniences (5)
(g) level of taxes and special assessments (5)
(h) appeal (10)

Although FHA directives insisted that no project should be insured that involved a high degree of risk with regard to any of the eight categories, "economic stability" and "protection from adverse influences" together counted for more than the other six combined. Both were interpreted in ways that were prejudicial against heterogeneous environments.[58] The 1938 *Underwriting Manual* taught that "crowded neighborhoods lessen desirability," and "older properties in a neighborhood have a tendency to accelerate the transition to lower-class occupancy."[59] Smoke and odor were considered "adverse influences," and appraisers were told to look carefully for any "inferior and non-productive characteristics of the areas surrounding the site."[60]

Obviously, prospective buyers could avoid many of these so-called undesirable features by locating in peripheral sections. In 1939, the Washington headquarters asked each of the fifty-odd regional FHA offices to send in the plans for six "typical American houses." The photographs and dimensions were then used for a National Archives exhibit. An analysis of the submissions clearly indicates that the ideal home was a bungalow or a colonial on an ample lot with a driveway and a garage.

In an attempt to standardize such ideal homes, the Federal Housing Administration set up minimum requirements for lot size, for setback from the street, for separation from adjacent structures, and even for the width of the house itself. While such requirements did provide air and light for new structures, they effectively eliminated whole sections of cities, such as the 16-foot-wide row houses of Baltimore, from eligibility for loan guarantees. Even apartment owners were encouraged to look to suburbia: "Under the best of conditions a rental development under the FHA program is a project set in what amounts to a privately owned and privately controlled park area."[61]

Reflecting the broad segregationist attitudes of a majority of the American people, the Federal Housing Administration was extraordinarily concerned with "inharmonious racial or nationality groups." Homeowners and financial institutions alike feared that an entire area could lose its investment value if rigid white-black separation was not maintained. Bluntly warning that "if a neighborhood is to retain stability, it is necessary that properties shall continue to be occupied by the same social and racial classes," the *Underwriting Manual* openly recommended "enforced zoning, subdivision regulations, and suitable restrictive covenants "that would be "superior to any mortgage."[62] Such convenants were a common method of prohibiting black occupancy until the U.S. Supreme Court ruled in 1948 (Shelley v. Kraemer) that they were "unenforceable as law and contrary to public policy." Even then, it was not until late 1949 that FHA announced that as of February 15, 1950, it would not insure any more mortgages on real estate subject to covenants. Although the press treated the FHA announcement as a major advancement in the field of racial justice,[63] former housing official Nathan Straus noted that "the new policy in fact served only to warn speculative builders who had not filed covenants of their rights to do so, and it gave them a convenient respite in which to file."[64]

In addition to recommending covenants, FHA compiled detailed reports and maps charting the present and most likely future residential locations of black families. In a March 1939 map of Brooklyn, for example, the presence of a single, nonwhite family on any block was sufficient to mark that entire block black. Similarly, very extensive maps of the District of Columbia depicted the spread of the black population and the percentage of dwelling units occupied by persons other than white.[65] As late as November 19, 1948, Assistant FHA Commissioner W. J. Lockwood could write that FHA "has never insured a housing project of mixed occupancy" because of the expectation that "such projects would probably in a short period of time become all-Negro or all-white."[66]

Occasionally, FHA racial decision were particularly bizarre and capricious. In the late 1930s, for example, as Detroit grew outward, white families began to settle near a black encalve near Eight Mile Road. By 1940, the blacks were surrounded, but neither they nor the whites could get FHA insurance because of the presence of an adjacent "inharmonious" racial group. So in 1941, an enterprising white developer built a concrete wall between the white and black areas. The FHA then took another look and approved mortgages on the white properties.[67]

One of the first persons to point a finger at FHA for discriminatory practices was Professor Charles Abrams. Writing in 1955, he said:

> A government offering such bounty to builders and lenders could have required compliance with a nondiscrimination policy. Or the agency could at least have pursued a course of evasion, or hidden behind the screen of local autonomy. Instead, FHA adopted a racial policy that could well have been culled from the Nuremberg laws. From its inception FHA set itself up as the protector of the all-white neighborhood. It sent its agents into the field to keep Negroes and other minorities from buying houses in white neighborhoods.[68]

The precise extent to which the agency discriminated against blacks and other minority groups is difficult to determine.

Although the Federal Housing Administration has always collected reams of data regarding the price, floor area, lot size, number of bathrooms, type of roof, and structural characteristics of the single-family homes it has insured, it has been quite secretive about the spatial distribution of those loans. For the period since 1942, the most detailed FHA statistics cannot be disaggregated below the county level.

Such data as are available indicate that neighborhood appraisals were influential in determining for FHA "where it would be reasonably safe to insure mortgages." Indeed, the preliminary examiner was specifically instructed to refer to the residential security maps in order "to segregate for rejection many of the applications involving locations not suitable for amortized mortgages." The result was a degree of suburban favoritism even greater than documentary analysis would have predicted. Of a sample of 241 new homes insured by FHA throughout metropolitan St. Louis between 1935 and 1939 (Figure 4.3), a full 220, or 91 percent, were located in the suburbs. Moreover, as Figure 4.3 indicates, more than half (135 or 241) of these homebuyers had lived in the city immediately prior to their new home purchases. That FHA was helping to denude St. Louis of its middle-class residents is illustrated by a comparison of the HOLC residential security map (Figure 4.1) with Figure 4.3. As might be expected, the new suburbanites were not being drawn from the slums or from rural areas, but from the better (A and B) areas of the central city.

A detailed analysis of two individual subdivisions in St. Louis County—Normandy and Affton—confirms the same point. Located just northwest of the city limits, Normandy (Figure 4.4) was made up in 1937 of new five- and six-room houses costing between $4,000 and $7,500. In 1937 and 1938, exactly 127 of these houses were sold under FHA guaranteed mortgages. Of the purchasers, 100 (78 percent) moved out from the city, mostly from the solid, well-established blocks between West Florrissant and Easton Streets

Affton was on the opposite, or southwest, edge of St. Louis, but it also was the scene of considerable residential construction in 1938 and 1939. Of 62 families purchasing FHA-

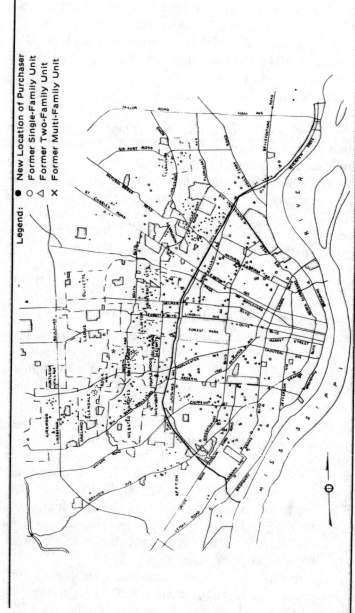

Legend:
● New Location of Purchaser
○ Former Single-Family Unit
△ Former Two-Family Unit
✗ Former Multi-Family Unit

Figure 4.3: Location and Type of Structure from Which Purchaser Moved to 241 New, Single-Family, FHA-Insured Homes Throughout Metropolitan St. Louis, 1935-1939

SOURCE: Record Group 31, National Archives.

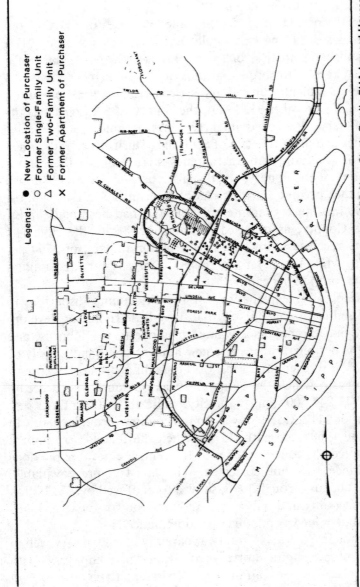

Legend:

● New Location of Purchaser

○ Former Single-Family Unit

△ Former Two-Family Unit

✕ Former Apartment of Purchaser

Figure 4.4: Location and Type of Structure from Which Purchasers Moved to 129 New, Single-Family FHA-Insured Houses in Normandy, St. Louis County, 1937-1938

SOURCE: Record Group 31, National Archives.

insured homes in Affton during those years (Figure 4.5), 55 were from the city of St. Louis. Most of them simply came out the four-lane Gravois Road from the southern part of the city to their new plots in the suburbs.

For the period since 1942, detailed analyses of FHA spatial patterns seem to be impossible. But a reconstruction of FHA unpublished statistics for the St. Louis area over a third of a century reveals the broad patterns of city-suburban activity. As Table 4.3 indicates, in the first sixteen years of FHA operation (through December 31, 1949), the county of St. Louis was the beneficiary of more than three times as much mortgage insurance activity as the city of St. Louis. During the 1950s, when tens of thousands of tract homes were built in the central portions of the county, the disparity between city and suburb assistance became startling. As of December 31, 1960, almost 63,000 insurance guarantees (Table 4.4) had been made in St. Louis County, as opposed to about 12,000 in the city. The county received more than half a billion dollars, or $794 per capita, while the city received less than one hundred million dollars, or $126 per capita.[69]

One possible explanation for the city-county disparities in these figures is that the city had very little room for development, that the populace wanted to move to the suburbs, and that the periphery was where new housing could most easily be built.[70] But even in terms of home improvement loans, a category in which the aging city was obviously more needy, only $43,844,500 went to the city, while about three times that much, or $112,315,798, went to the country through 1960.[71] In the late 1960s and early 1970s, when the federal government attempted to redirect monies to the central cities, the previous imbalance was not corrected. The latest figures available, which take us through 1976, show a total of well over $1 billion for the county and only about $300 million for the city. Thus, the suburbs have continued their dominance.[72]

Although St. Louis County apparently has done very well in terms of per-capita mortgage insurance in comparison with other areas of the nation, the Mississippi River was not an

(text continued on p. 106)

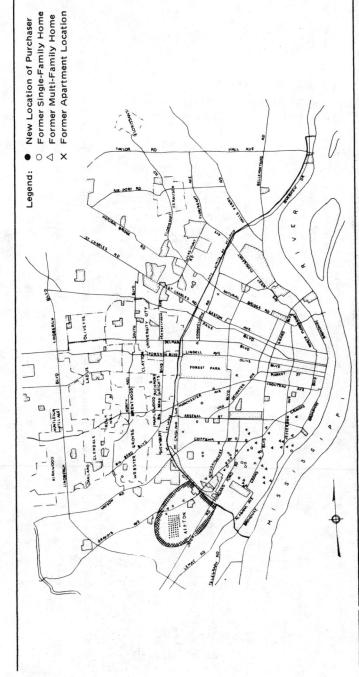

Legend:
- ● New Location of Purchaser
- ○ Former Single-Family Home
- △ Former Multi-Family Home
- X Former Apartment Location

Figure 4.5: Location and Type of Structure from Which Purchasers Moved to 62 New, Single-Family, FHA-Insured Homes in Affton, St. Louis County, 1938-1939

SOURCE: Record Group 31, National Archives.

TABLE 4.3 Cumulative FHA Home Mortgage Activities for Ten Selected Counties, 1934-1949

Jurisdiction	Cumulative Number of Home Mortgages, 1934-1949	Cumulative Amount of Home Mortgages 1934-1949	Per-Capita Amount of Home Mortgages as of January 1950
St. Louis County, Missouri	22,795	$133,795,533	$329
Nassau County, New York	31,165	191,510,973	285
Montgomery County, Maryland	5,735	39,454,100	240
Fairfax County, Virginia	2,697	18,289,099	186
Prince Georges County, Maryland	5,616	31,473,217	162
District of Columbia	5,875	40,649,862	51
St. Louis City, Missouri	6,695	38,883,972	45
Kings County (Brooklyn), New York	8,944	52,131,887	19
Hudson County, New Jersey	847	4,905,520	8
Bronx County, New York	1,054	6,361,293	4

SOURCE: These calculations are based on unpublished statistics available in loose-leaf binders in the Single Family Insured Branch of the Management Information Systems Division of the Federal Housing Administration.

NOTE: The per-capita amount was derived by dividing the cumulative amount of home mortgages through December 31, 1949, by the total population as of the 1950 census.

TABLE 4.4   Cumulative FHA Home Mortgage Activities and Per-Capita Figures for Ten Selected Counties, 1934-1960

Jurisdiction	Cumulative Number of Home Mortgages, 1934-1960	Cumulative Amount of Home Mortgages 1934-1960	Per-Capita Amount of Home Mortgages as of January 1961
St. Louis County, Missouri	62,772	$558,913,633	$794
Fairfax County, Virginia	14,687	190,718,799	730
Nassau County, New York	87,183	781,378,559	601
Montgomery County, Maryland	14,702	159,246,550	467
Prince Georges, County, Maryland	15,043	144,481,817	404
St. Louis City, Missouri	12,166	94,173,422	126
District of Columbia	8,038	66,144,612	87
Kings County (Brooklyn), New York	15,438	140,330,137	53
Hudson County, New Jersey	1,056	7,263,320	12
Bronx County, New York	1,641	14,279,243	10

SOURCE: These calculations are based on unpublished statistics available in the Single Family Insured Branch of the Management Information Systems Division of the Federal Housing Administration.

NOTE: The per-capita amount was derived by dividing the cumulative amount of home mortgages by the 1960 population.

isolated case of FHA suburban favoritism. In Essex County, as
Figure 4.6 indicates, FHA commitments went in overwhelming
proportion to Newark's suburbs.[73] And in neighboring Hudson
County, New Jersey, residents received only $12 of mortgage
insurance per capita through 1960, the second lowest county
total in the nation after the Bronx (Table 4.4).[74]

The New Jersey data reveal that the most favored areas for
FHA mortgage insurance were not the wealthiest towns.
Rather, the most likely areas for FHA activity were those rated
B on the residential security maps. In 1936, 65 percent of new
housing units in suburban Livingston were accepted for insurance; for Caldwell and Irvington, also solidly middle-class,
the percentages were 59 and 42 respectively.[75] In elite districts
like South Orange, Glen Ridge, Milburn, and Maplewood,
however, the FHA assistance rates were about as low as they
were for Newark, or less than 25 percent. Presumably this
occurred because the housing available in the so-called best
sections was beyond the allowable price limits for FHA mortgage insurance, and also because persons who could afford to
live in such posh neighborhoods did not require government
financing.[76]

Even in the nation's capital, the outlying areas were considered more appropriate for federal assistance than older
neighborhoods. FHA commitments at the beginning of 1937 in
the District of Columbia were heavily concentrated in two
peripheral areas: (1) between the U.S. Soldiers' Home and
Walter Reed Hospital in white and prosperous northwest
Washington, and (2) between Rock Creek Park and Connecticut Avenue, also in Northwest Washington. Very few
mortgage guarantees were issued in the predominantly black
central and southeastern sections of the district. More important, at least two-thirds of the FHA commitments in the
metropolitan area were located in the suburbs—especially in
Arlington and Alexandria in Virginia and in Silver Spring,
Takoma Park, Bethesda, Chevy Chase, University Park,

Figure 4.6: New Home Mortgages Accepted for Insurance by the Federal Housing Administration in Essex County, New Jersey, in 1936 (each dot represents one mortgage accepted for insurance)

SOURCE: Record Group 31, National Archives.

Westmoreland Hills, and West Haven in Maryland. Perhaps this was but a reflection of the 1939 FHA prediction:

> It should be noted in this connection that the "filtering-up" process, and the tendency of Negroes to congregate in the District, taken together, logically point to a situation where eventually the District will be populated by Negroes and the suburban areas in Maryland and Virginia by white families.[77]

Following a segregationist policy for at least the next twenty years, the FHA did its part to see that the prophecy came true; through the end of 1960, as Table 4.4 indicates, the suburban counties had received more than seven times as much mortgage insurance as the District.

For its part, the FHA usually responded that it was not created to help cities but to revive homebuilding and to stimulate homeownership. And it concentrated on convincing both Congress and the public that it was, as its first administrator, James Moffett, remarked, "a conservative business operation."[78] The agency emphasized its concern over sound loans, no higher than the value of the assets and the repayment ability of the borrower would support, and its ability to make a small profit for the federal government.

But the FHA also helped to turn the building industry against the minority and inner-city housing market, and its policies supported the income and racial segregation of most suburbs. Whole areas of cities were declared ineligible for loan guarantees;[79] as late as 1966, for example, the FHA did not have a mortgage on a single home in Camden, New Jersey, a declining industrial city.[80]

Despite the fact that the government's leading housing agency was openly exhorting segregation, throughout the first thirty years of its operation, very few voices were raised against FHA's redlining practices. Between 1943 and 1945, Harland Bartholomew and Associates prepared a series of reports as a master plan for Dallas. The firm criticized FHA for building "nearly all housing" in the suburbs, and argued that "this policy has hastened the process or urban decentralization immeasurably."[81] And in 1955 Charles Abrams objected to FHA policies that had "succeeded in modifying legal practices so that the common form of deed included the racial covenant."[82]

Not until the civil rights movement of the 1960s did community groups and scholars become convinced that redlining and disinvestment were a major cause of neighborhood decline[83] and that home improvement loans were the "lifeblood of housing."[84] In 1967, Martin Nolan summed up the indictment against FHA by asserting, "The imbalance against poor people and in favor of middle-income homeowners is so staggering that it makes all inquiries into the pathology of slums seem redundant."[85] In the following year, Senator Paul Douglas of Illinois reported for the National Commission on Urban Problems on the role of the federal government in home finance:

> The poor and those on the fringes of poverty have been almost completely excluded. These and the lower middle class, together constituting the 40 percent of the population whose housing needs are greatest, received only 11 percent of FHA mortgages .... Even middle-class residential districts in the central cities were suspect, since there was always the prospect that they, too, might turn as Negroes and poor whites continued to pour into the cities, and as middle and upper-middle-income whites continued to move out.[86]

Moreover, as Jane Jacobs has said, "Credit blacklisting maps are accurate prophecies because they are self-fulling prophecies."[87]

The main beneficiary of the $119 billion in FHA mortgage insurance issued in the first four decades of FHA operation was suburbia, where approximately half of all housing could claim FHA or VA financing in the 1950s and 1960s. In the process, the American suburb was transformed from a rich person's preserve into the normal expectation of the middle class.[88]

THE PUBLIC HOUSING PROGRAM

The long-term, low-interest mortgage was not the only federal housing program to benefit the suburbs at the expense of the cities. More controversial was the attempt to meet the housing needs of the poor. Prior to the 1930s, housing reform in the United States meant the improvement of slum conditions through the establishment of minimum standards of ventilation, density, and sanitation. New York City's pioneering codes of 1867, 1879, and 1901, each of which established progressively higher legal requirements for dwelling units, were illustrative of this trend, as were the ideas of the nation's preeminent nineteenth-century housing reformer, Lawrence Veiller, who opposed government action beyond the enforcement of the law.[89]

Soon after World War I, however, the widely traveled wife of a naval officer became the first American to support effectively "positive" rather than "negative" housing reform. Edith Elmer Wood maintained that social behavior was conditioned by housing and that government action to replace the slums would reduce crime and delinquency and result in better citizenship and lower community welfare costs. Having witnessed the initiatives European nations were taking to shelter their inhabitants, she subsequently devoted herself to the campaign for actual government construction of dwelling units. In books such as *The Housing of the Unskilled Wage Earner* (New York, 1919), she argued that private philanthropy was not the solution to the housing problem and that restrictive building codes simply raised the rent levels of tenements while doing nothing at all to increase the supply.[90] Prior to the New Deal, however, only the states of New York and North Dakota accepted the provision of housing as even a limited responsibility.[91]

In an important reversal of traditional federal policy, the administration of Franklin D. Roosevelt initiated its own construction program. The direct involvement of Uncle Sam began with the passage of the National Industrial Recovery Act during the First Hundred Days of 1933. The legislation had four purposes: to increase employment, to improve housing for the poor, to demonstrate to private industry the feasibility of large-scale community planning efforts, and to eradicate and rehabilitate slum areas in order "to check the exodus to the outer limits of cities with consequent costly utility extensions and leaving the centrally located areas unable to pay their way."[92] The first purpose was the most important. The Congress wanted to create jobs, not housing.

This 1933 housing law authorized the Public Works Administration to accomplish these purposes through three mechanisms. First, the PWA Housing Division could lend money to private, limited-dividend corporations interested in slum clearance. Second, grants and loans could be made available to public authorities for the same purpose. Third, and most significant, the Housing Division was empowered to buy, condemn, sell, or lease property for developing new projects itself.[93]

Although PWA Administrator Harold Ickes bluntly complained that "American cities cannot produce a single instance in which slums have been cleared and new dwellings built to rehouse the dispossessed occupants by private enterprise operating on a commercial basis,[94] the PWA attempted to place the emphasis of the program on private development encouraged by federal loans. Predictably, however, only seven of five hundred limited-dividend corporations that applied had sufficient equity to qualify for the program, and those seven seemed primarily anxious to sell land to the government at inflated prices.[95] Unable to rebuild the slums through this provision of the law, the Housing Division terminated its limited-dividend program in mid-1934 and turned to its other alternatives.[96]

The ability of the PWA to work with local authorities proved of limited usefulness. In 1933, no state or locality had the legal authority to engage in slum clearance projects; as late as 1937 only New York, Ohio, Michigan, and South Carolina had passed the required enabling legislation. Thus, the Housing Division was forced to construct its own low-income housing projects on land acquired by condemnation or purchase. Between 1934 and 1937, when the Housing Division of PWA was replaced by the U.S. Housing Authority (USHA), the government began only 21,000 units.[97]

Here also the PWA encountered obstacles. In a landmark decision handed down in January 1935, Federal Judge Charles I. Dawson of Kentucky ruled that acquiring land for public housing in Louisville by condemnation (eminent domain) was not constitutional and that the PWA could not therefore exercise this power. In the words of Judge Dawson, low-cost housing

> is certainly not a public use, in the sense that the property is proposed to be used by the federal government for performing any of the legitimate functions of the Government itself. Surely it is not a governmental function to construct buildings in a state for the purpose of selling or leasing them to private citizens for occupancy as houses.[98]

Although lawyers for the PWA appealed Dawson's ruling all the way to the U.S. Supreme Court, they withdrew the appeal a few hours before oral arguments were to be heard.[99] Thereafter, the Housing Division complied with the decision by purchasing land through negotiated sales at higher prices than condemnation would have incurred. Costs also rose because this first New Deal housing effort was a hastily designed program primarily designed to put men back to work. Construction was a means of employment more than an end in itself. Thus, by the time the PWA erected apartments or houses, the minimum rents that had to be charged precluded occupancy by the urban poor. Elements within the Roosevelt administration suggested that the projects be operated at a loss, but the proposal was shelved when the Comptroller General ruled that there was no law to authorized such a subsidy.[100]

Well aware that adverse judicial decisions and escalating costs would effectively cripple the PWA housing program, Senator Robert F. Wagner of New York and Representative Henry Ellenbogen of Pennsylvania introduced new legislation in 1935 to create a permanent public housing agency. Initially, President Roosevelt gave the Wagner-Ellenbogen measure only lukewarm support, and it died in the House Banking and Currency Committee.[101] The platform of the Democratic party in 1936 contained only a weak plank endorsing federal involvement in housing, and the issue was of minor significance in the election.

The unprecedented magnitude of reelection landslide, however, made the President feel safe in his advocacy of a stronger Washington role in housing. In his second inaugural address, he came out dramatically on the issue:

> But here is the challenge to our democracy. In this nation I see tens of millions of its citizens—a substantial part of its whole population—who at this very moment are denied the greater part of what the very lowest standards of today call the necessities of life.... I see one-third of a nation ill-housed, ill-clad, ill-nourished.[102]

Although the President was personally more interested in back-to-the-land movements than in public housing, and although he shared the confusion of slum clearance with improving conditions for the poor that was so common in the 1930s and beyond, Roosevelt did give public support to public housing in the spring of 1937. Within a matter of months, the U.S. Housing Act (USHA) had passed the Senate by 64 to 16 and the House by 275 to 86. It was signed on September 1, 1937. Long-time reformer Catherine Bauer called it "a radical piece of legislation," and the New York *Times* added: "With the President's signature the Wagner Steagall bill becomes law and at last America makes a real start toward wiping out its city slums."[103]

The legislation empowered the USHA to develop public projects by funding duly constituted local housing agencies. The USHA was to funnel this money to municipalities through two mechanisms: first, by lending up to 90 percent of the capital costs of a project to local officials,[104] and second, by subsidizing construction and maintenance costs. So enthusiastic was President Roosevelt that when work began on the first five projects under the new procedures on March 17, 1938, he wrote to Nathan Straus, his chief housing official: "Today marks the beginning of a new era in the economic and social life of America. Today, we are launching an attack on the slums of this country which must go forward until every American family has a decent home."[105]

On one level, public housing was a resounding success. By the end of 1938, 221 local authorities had been established in the thirty-three states that had passed enabling legislation. By the end of 1962, more than two million people lived in the half million units built under various public housing programs. If the quality and design of the projects frequently invited derision, they were nevertheless superior to the delapidated structures they replaced.[106]

On another level, however, public housing did not fulfill the expectations of its supporters. There never was enough of it. As late as 1979, fewer than 2 percent of Americans dwelled in such housing, as compared with more than a quarter of the British population. And the problem was a shortage of funding, not a shortage of need. But because the real purpose of the 1937 law was to alleviate "present and recurring unemployment," as soon as the nation entered wartime and postwar housing booms, the public sector was relegated to a low priority.[107]

Of particular importance to the spatial distribution of the limited amount of public housing that was built was the decentralized nature of the program. In view of Judge Dawson's ruling and of widespread opinion that federal use of the power of eminent domain for housing construction was unconstitutional, Senator Wagner's bill created the USHA as a "low-rent housing and slum clearance measure . . . drawing its strength from *local initiative and responsibility*" (emphasis mine). It required that any city desiring public housing had to provide tax exemptions for the project and had to create a local housing agency. Thus, every community had to make its own decision as to whether or not a need existed; the resulting application for federally subsidized housing had to be a *voluntary* action.

That distinction was critical. A suburb that did not wish to tarnish its exclusive image by having public housing within its precincts could simply refuse to create a housing agency. No local housing authority from another jurisdiction and no national official could force it to do otherwise.[108] Needless to say, hundreds of suburbs throughout the United States have yet to apply.[109] Meanwhile, Newark, New Jersey, one of the most troubled of American cities, has more units of public housing per capita than any other community in the nation.

A second feature of the legislation that tended to concentrate public housing in the center rather than on the periphery was the fact that housing authorities typically consisted of prominent citizens who were more anxious to clear slums and protect real-estate values than they were to rehouse the poor. As John F. Bauman has clearly demonstrated in a study of Philadelphia, the public housing authority was especially anxious to boost the sagging urban tax structure, to halt the spread of blight, and to raise property values.[110]

Finally, there was a requirement that one slum unit be eliminated for every unit of public housing erected. Thus, only localities with significant numbers of inadequate dwellings could receive assistance. The following exchange between Representative Kunkel of Pennsylvania and Commissioner Egan of the Housing Authority underscores the point:

> *Mr. Kunkel:* Under this program, no area in which there is no substandard housing would be eligible for any public housing. Is that correct?
>
> *Commissioner Egan:* That is correct. If there were no slums in that locality, regardless of how acute the housing shortage was, and if we knew we could not get the equivalent elimination required by the act, we could not go in there.[111]

Even the most progressive congressional leaders accepted such limitations out of concern that to do otherwise would imperil passage. In 1949, when a major new public housing act was being debated, Republican opponents proposed an amendment that forbade any form of racial or ethnic discrimination in all public housing units. This placed many of the bill's supporters in a dilemma. If the amendment succeeded, the southern senators would be certain to vote against the entire bill, thus ensuring its defeat. Yet, northern liberals were not inclined to vote against an amendment promising racial justice.

Senator Paul Douglas urged liberals to put aside their principles temporarily in order to give the measure a chance. He told his colleagues that the amendment "necessarily creates a sharp conflict within the hearts of all of us who want on one hand to clear the slums and to provide housing for the slum dwellers and who, at the same time, feel very keenly that we should not treat any race as second-class citizens."[112] He went on to say,

> I am ready to appeal to history and to time that it is in the best interests of the Negro race that we carry through the housing program as planned, rather than put in the bill an amendment which will inevitably defeat it, and defeat all hopes for rehousing 4,000,000 persons.[113]

Because public housing was confined to existing slums and because determination of need and site selection were left up to the localities, the program further concentrated the poor in the city and reinforced the image of suburbia as a place of refuge from the social pathologies of the poor.[114] There was even a concentration within particular parts of cities. In Chicago, for example, 150,000 persons lived in low-income public housing in 1978. A few scattered projects were in marginal white neighborhoods, such as the late Mayor Richard Daley's own Bridgeport. The occupants were predominantly white. The other 95 percent of Chicago's public housing, however, was dumped into the most poverty-impacted black ghettos in the city.[115]

By the late 1960s, social commentators like Herbert Gans and Lee Rainwater were describing public housing as "federally built and supported slums." The need remained, and the waiting time for admission to many projects was measured in years. But public housing was often architecturally bland and sometimes physically dangerous. Because it was the "dumping ground for the poor," it served the purposes of social control well enough. But an occasional disaster, such as the notorious Pruitt-Igoe project of St. Louis, demonstrated that a minimal level of planning and amenities were necessary even for the most disadvantaged.

CONCLUSION

In assessing the effect of federal policies toward housing on the movement of white, middle-class families away from older, industrial cities and toward newer, suburban areas, any analysis must take cognizance of the following points:

First and most obviously, it is hazardous to condemn a government for adopting housing policies in accord with the preferences of a majority of its citizens. As novelist Anthony Trollope put it in 1867, "It is a very comfortable thing to stand on your own ground. Land is about the only thing that can't fly away." For more than a century Americans have had a strong affinity for a detached home on a private lot. Obviously, some popular measures, such as health insurance and gun control, are not adopted because of powerful special interest lobbies. But suburbanization was not willed upon an innocent peasantry. Without a substantial amount of encouragement from the mainstream of public opinion, the bureaucrats would never have been able to push their projects as far as they did. In fact, suburbanization was an ideal government policy because it met the needs of both citizens and business interests and because it got the politicians votes. It is a simple fact that homeownership introduced equity into the estates of over 35 million families between 1933 and 1978. The tract houses they often bought may have been regarded as hopeless by architectural purists, but they were a lot less dreary to the people who raised families there and then sold to new families at a profit.

Second, to the extent that Washington adopted a pro-suburban housing policy, it simply followed in the well-worn path of state and local governments. The method of opening streets in urban America is instructive in this regard. Before the Civil War, streets were improved when owners of a certain percentage (usually three-fourths) of the property facing the right-of-way petitioned the city government to do so. To finance such improvements, property owners "abutting and directly affected" paid special assessments to meet the costs of paving. Because the owners would presumably benefit from the increased value of their land after the street was opened, the system had a certain logic and justification. After mid-century, however, a second method of financing became more common, one that passed the cost of peripheral street improvements on to the municipality as a whole. This method naturally appealed to real-estate speculators and builders who favored suburban development.[116]

Tenement house laws are another example. The 1867, 1879, and 1901 New York City ordinances, which influenced similar measures across the nation, did not so much alleviate conditions in immigrant neighborhoods as ensure that the worst abuses would not be reproduced in the newly developing sections of Brooklyn and the Bronx. New requirements dealing with light and ventilation could not easily be applied retroactively to the crowded buildings of the Lower East Side and East Harlem, but they could be enforced with regard to any new construction. Similarly, methods of constructing schools and sewers exhibited the same pattern of creating the best environment on the edges and, if necessary, paying for it by taxing the entire city.[117]

Third, the federal government, and especially the FHA, established minimum standards for home construction that became almost universal in the industry. In recent years, the largest private contractors have built all their new homes to meet FHA standards, even though financing has often been arranged without any FHA aid. This has occurred because many potential purchasers will not consider a home that cannot get FHA approval.[118]

Thus, the FHA was not the sine qua non in the mushrooming of the suburbs; neither did the ghetto or the slum originate in the public housing effort. But in a critical period of American history, the national government put its seal of approval on ethnic and racial discrimination and developed policies that had the result, and I believe the intention, of the practical abandonment of large sections of older, industrial cities. Washington actions were later picked up by private interests, so that it is now the banks and savings and loan associations that are most guilty of denying mortgages "solely because of the geographical location of the property."

Political commentators have often remarked that federal policies are often at cross purposes with one another, that the left hand seems never to know what the right hand is doing. Perhaps in an organization as immense as the U.S. government it would be impossible to have a single, consistent policy in dealing with basic national issues. But however confused the situation appears, the basic direction of federal housing policies has been toward the concentration of the poor and the suburban dispersal of the better-off.[119]

St. Louis illustrates the dilemma of many cities. Partly as a result of federal housing policies that have enabled the white, middle-class population to settle in the county, the city of St. Louis has slid steadily downhill, and it is now a premier example of urban abandonment. Once the fourth largest city in America, the "gateway to the west" is now the twenty-third, a ghost of its former self.[120] In 1940, it contained 816,048 inhabitants; the 1980 census will show about 550,000. Many of its old neighborhoods have become dispiriting collections of burned-out buildings, eviscerated homes, and vacant lots. There is an eerie remoteness to the pockmarked streets, even though the drone of traffic on the nearby interstate highways is constant. The air is polluted, the sidewalks are filthy, the juvenile crime rate is horrendous, and industry is languishing. Grimy warehouses and aging loft factories are landscaped by weed-grown lots adjoining half used rail yards. Like an elderly couple no longer sure of their purpose in life after their children have moved away, these neighborhoods face an undirected future.[121]

A particularly telling statistic is that, after Chicago, St. Louis is the nation's leading exporter of used bricks. Piled beside the railroad tracks that hug the Mississippi River, the great stacks of weathered bricks are destined to become parts of restoration projects in Atlanta or patios in Houston. It is the supreme indignity. Having lost more than 300 factories in the past decade to the Sunbelt, St. Louis itself is now being carted away.

The situation in the Mississippi River metropolis is more serious than that in most other cities, but the same broad patterns of downtown decline, inner-city deterioration, and exurban development so evident in St. Louis are actually typical of the large population centers of the United States. This same result might have been achieved in the absence of all government intervention, but the simple fact is that the various federal policies toward housing have had essentially the same effect from Portland to Portsmouth. The poor in America have not shared in the postwar real-estate boom, in most of the major highway improvements, in property and income-tax writeoffs, and in mortgage insurance programs. At the same time, the much-needed public housing effort has been consistently underfunded and has served to concentrate and control the disadvantaged as much as to improve their standard of living.

In the past few years, following a trend begun in Boston, San Francisco, New York, Washington, Charleston, and Savannah, some of the old neighborhoods of St. Louis have experienced an influx of investors and residents. As older housing regains its cachet, we can expect the federal government to shift its housing policies in the direction of more assistance for renovation, for rehabilitation, for inner-city beautification, and for mass transit. After all, as the South African experience indicates, social control does not require that ghettos invariably be located at the center. If the affluent ever choose to return to the city, the government will help finance it, and the poor will pay the price.

NOTES

1. This statement, which expresses a widely shared view, was made by political scientist Ralph A. Rossum. *Memphis Press-Scimitar*, January 6, 1977.

2. Seward H. Mott, "The Case for Fringe Locations," *Planners Journal* 5 (March-June 1939), 38.

3. The tax code gives older cities a built-in handicap that will continue to work against them no matter how low their fortunes sink. For example, the extension of the investment tax credit for business machinery purposes to outlays on new construction could make a large part of the industrial plant of the Middle West and Northeast obsolete overnight. *Business Week*, December 19, 1977, 87-88.

4. Senator Daniel P. Moynihan, a former White House urban affairs adviser, estimated that New York State lost $10.6 billion in the federal exchange in 1976.

5. Interstate highways have provided benefits to thinly populated states as well as to suburbs. Between 1957 and 1972, according to Roger Vaughan, Montana received $2.44 worth of highway for each dollar it paid into the highway trust fund. Wyoming received $2.71 and Nevada $1.98, while Massachusetts received only $.77, New Jersey $.66, and New York $.80. *Business Week*, December 19, 1977, 88.

6. This is because the emphasis of sewer and water aid is on *new* construction. New York City's immense, century-old water system, in need of billions of dollars worth of repairs, is not eligible for the large grants.

7. For a conspiratorial view of the state, see Peter Marcuse, "The Myth of the Benevolent State: Notes Toward a Theory of Housing Conflict (unpublished, 1978).

8. An example would be a New York City building regulation of 1766 creating a fire zone where houses had to be made of stone or brick and roofs of tile or slate.

9. The best full-scale studies of federal housing programs are: Mark I. Gelfand, *A Nation of Cities: The Federal Government and Urban America, 1933-1965* (New York, 1975); Henry Aaron, *Shelter and Subsidies: Who Benefits from Federal Housing Policies* (Washington, Brookings Institution, 1972); William L.C. Wheaton, "The Evolution of Federal Housing Programs" (Ph.D. dissertation, University of Chicago, 1953); and Paul F. Wendt, *Housing Policy: The Search for Solutions* (Berkeley, 1962), 142-273. On the 1970s, see A. Naparstek and G. Cincotta, *Urban Disinvestment: New Implications for Community Organizations, Research, and Public Policy* (Washington, National Center for Urban Ethnic Affairs and the National Training and Information Center, 1976); and Calvin Bradford, "Financing Home Ownership: The Federal Role in Neighborhood Decline," *Urban Affairs Quarterly* 14 (March 1979), 313-335.

10. In February 1919, the Department of Labor set up an "Own Your Own Home" section in its Division of Public Works and Construction Development. Its purpose was to publicize the housing campaign of the National Association of Real Estate Boards.

11. During the 1920s, under the leadership of Governor Alfred E. Smith, New York State pioneered in the housing field by inducing private corporations to construct cooperative apartments and projects in return for exemption from state and local taxes. About 6,000 units were built under this program.

12. Edith Elmer Wood, *Recent Trends in American Housing* (New York, 1931), 12-20. See also Wallace F. Smith, *Housing: The Social and Economic Elements* (Berkeley, 1971).

13. Semer and Zimmerman, *Evolution of Federal Legislative Policy in Housing: A Report to HUD* (Consultant's Report dated June 30, 1973), III-1 through III-15.

14. Hoover was not unusual. President Calvin Coolidge said, "No greater contribution could be made to the stability of the Nation and the advancement of its ideals, than to make it a nation of home-owning families." Quoted in Glenn H. Beyer, *Housing and Society* (New York, 1965), 249.

15. Quoted in Lyle Woodyatt, "The Origins and Evolution of the New Deal Housing Program" (Ph.D. dissertation, Washington University, 1968).

16. Amortization refers to the repayment of the principal in full by the expiration date of the mortgage. Prior to 1932, many loans required only the payment of the interest, with the entire original amount being due at the expiration of the loan.

17. *Final Report of the Committee on Large-Scale Operations, The President's Conference on Home Building and Home Ownership* (Washington, Government Printing Office, 1932), 24.

18. *Emergency Relief and Construction Act*, (1932), Public Law 302, 72nd Congress, Title II, Section 201.

19. The best study of this subject is Joseph L. Arnold, *The New Deal in the Suburbs: A History of the Greenbelt Town Program* (Columbus, Ohio, 1971).

20. This refers to the point made earlier that the price of real estate *declined* between the late 1920s and the early 1930s.

21. The vote was 383 to 4 in the House; it passed without record vote in the Senate. C. Lowell Harriss, *History and Policies of the Home Owners' Loan Corporation* (New York: National Bureau of Economic Research, 1951), 11.

22. The HOLC's active lending program ended in 1936, and it was liquidated in the spring of 1951. It was administered by, and represented a major part of, the Federal Home Loan Bank Board.

23. Harris, *Home Owners' Loan Corporation*, 2.

24. "Redlining" is the term used to describe the arbitrary decisions of financial institutions not to lend in certain neighborhoods because of the age of the housing, the mixture of uses, the race of the inhabitants, or the proximity of the poor.

25. These comments are taken from the questionnaires that ordinarily accompanied the residential security maps. In this instance, they are from the St. Louis, Newark, and New York metropolitan areas.

26. Even the possibility of change was sufficient to lower a rating. In Westchester County, New York, the city of Mount Vernon's best neighborhoods were described as "well maintained and evidence pride of ownership. Nevertheless, the security grade was only B because of "the possible influx of less desirable elements from the Bronx." The maps and the appraisals are in Record Group 195 of the National Archives.

27. The information in this paragraph is taken from Bradford, "Financing Home Ownership," 319-325.

28. Hoyt was a professor of real estate and Park a professor of sociology. Hoyt's theories were later published as *The Structure and Growth of Residential Neighborhoods in American Cities* (Washington, 1939).

29. Although other scholars have noted the discriminatory result of federal housing policies, this investigation is the first to make systematic or any other use of the residential security maps and the detailed reports which supported them.

30. Because federal housing data are usually reported by county, I selected St. Louis and Newark because they are made up of separate and distinct counties.

31. "Highly restricted" was a shorthand way to indicate that Jewish residents were not welcome.

32. Lincoln Terrace was just east of Brentwood and south of Richmond Heights.

33. St. Louis City and County were legally separated in 1876 and have remained distinct since that time.

34. Those few St. Louis City neighborhoods that were rated First or Second tended to be located near attractive open spaces like Forest Park, Francis Park, or Carondolet Park.

35. *Metropolitan St. Louis: Summary of an Economic, Real Estate and Mortgage Finance Survey* (Washington, Division of Research and Statistics, Federal Home Loan Bank Board, 1942), especially 4, 11, and 12.

36. In this instance, as in many others, the federal appraisers were more aware of suburbanizing trends than were many other pre-World War II observers.

37. A good description of Newark in this period is in Philip Roth's novel, *Goodbye Columbus*.

38. This statement appeared on all of the HOLC evaluation ratings.

39. *Metropolitan St. Louis*, 16.

40. There is some evidence that the FHA simply used the HOLC residential security maps, but I have been unable to demonstrate this conclusively.

41. The influence went both ways because, as mentioned previously, private realtors and bankers helped to draw up the maps in the first place.

42. Because the VA very largely followed FHA procedures and attitudes and was not itself on "the cutting edge of housing policy," I shall discuss their accomplishments as a single effort. Bradford, "Financing Home Ownership," p. 332. For the economic significance of these agencies and of later public housing efforts, see Lawrence N. Bloomberg, "The Housing Problem: Long-Run Effects of Government Housing Programs," *American Economic Review* 41 (May 1951), 589-590.

43. Marion Clawson, *Suburban Land Conversion in the United States: An Economic and Governmental Process* (Baltimore, 1971), 80-91.

44. Aaron, *Shelter and Subsidies,* 76.

45. A major reason that long-term mortgage arrangements were not common prior to the 1930s was an 1864 amendment to the 1863 National Bank Act. It prohibited nationally chartered banks from making direct loans for real-estate transactions.

46. Although the full faith and credit of the United States stood behind the FHA insurance obligation, a special premium from homeowners (normally about .5 percent) almost always equaled or exceeded FHA expenses. As a result, FHA returned a profit to the government through its first twenty-five years. *Twenty-fifth Annual Report of the Federal Housing Administration for the Year Ending December 31, 1958.* (Washington, Federal Housing Administration, 1959), Section 1.

47. Only Iceland, Australia, and New Zealand, among all industrialized countries, exceeded the United States homeowner rate in 1972. Jim Kemeny, "Forms of Tenure and Social Structure," *British Journal of Sociology* 29 (March 1978), 43.

48. Interview with Martin Winter, April 19, 1977, New York City.

49. The VA program, which began in 1944, was even more generous than the FHA. The VA very largely followed FHA procedures and attitudes and was not itself on "the cutting edges of housing policy." Interview with Martin Winter, April 19, 1977, New York City.

50. In practice, FHA programs have operated largely in new residential developments on the edge of metropolitan areas, to the neglect of core cities, small towns, and rural areas.

51. There are important exceptions to this generalization. The notorious 608 program of the late 1940s), which Martin Winter called "the most colossal fraud of all time," offered inducements for multifamily construction that were in many instances much more lucrative than the single-family opportunities that FHA and VA presented. The government often lent builders as much as 30 percent more than the cost of construction, which meant that they could be very large developers without risking any

money of their own. This method is recounted in Charles Abrams, *The City Is the Frontier* (New York, 1965), 87-92.

52. Title I of the 1934 National Housing Act was for "Housing Renovation and Modernization." It insured financial institutions against losses sustained from loans for alterations, repairs, and improvements on real property. Its ineffectiveness was admitted in 1954 by Albert M. Cole, then the Administrator of the Housing and Home Finance Agency (of which FHA was one part), when he noted that Title I "is of limited assistance to families of modest income who need to finance home improvements." Committee on Banking and Currency, *Housing Act of 1954, Hearings*, 2 vols. (Washington, 1954), 52. See also President's Advisory Committee on Government Housing Policies and Programs, *Recommendations on Government Housing Policies and Programs* (Washington, 1953), 73; and Federal Housing Administration, *Remodel—Repair—Repay with FHA* (Washington: FHA, 1955), 1 and 6.

53. The National Housing Act Amendments of 1938 made mortgage insurance available for rental housing for the first time on a substantial basis, but as noted above, not much insurance was issued.

54. The original statute made appraisals necessary because maximum mortgage amounts were related to "appraised values."

55. Sellers had a huge stake in the appraisal decision on both new and existing homes because the FHA appraisal price often turned out to be the selling price. In the early 1970s, a substantial scandal occurred in New York, Detroit, and other cities when appraisers overvalued renovated homes and then guaranteed the mortgages which were secured by low down payments. When the shoddy repair work became evident and the houses often started to crumble, FHA was left holding the bag. See, for example, New York *Times*, March 20, 1972; June 28, 1972; and January 3, 1977.

56. *FHA Underwriting Manual* (Washington: FHA, 1947), Section 1301.

57. These are the categories established by the 1938 *Underwriting Manual*. By 1958, the titles of the first two categories had been changed to "Physical and Social Attractiveness" and "Protection Against Inharmonious Land Uses."

58. Prior to 1938, the *Underwriting Manual* was available only in typescript. A second edition was issued in 1947.

59. It is important to remember that in the 1930s the general direction of real-estate values was *down*. This deflation, coupled with the view that there was a cycle of occupancy from the affluent to the poor, naturally made appraisers especially cautious. They knew that houses very rarely sold for their assessed valuations and that the question was not whether a house or neighborhood would decline, but by how much and by when.

60. These comments are taken from Sections 1303 through 1316 of the 1938 *Underwriting Manual*.

61. Federal Housing Administration, *Rental Housing as Investment* (Washington: FHA, 1938), 30.

62. Neither the 1938 nor the 1947 *FHA Underwriting Manual* specifically endorsed "racial" covenants, but in the context of other directives and comments, there can be little doubt but that racially restrictive covenants were deemed desirable by FHA appraisers. Such covenants, which were part of the deed, required that no person of African descent ever be allowed to live on the property except as domestic servants or laborers.

63. Between 1934 and 1950 there was no FHA concern with equal opportunity in housing, and race was considered only to the extent that changing neighborhood composition would cause land values to fall.

64. Nathan Straus, *Two-Thirds of a Nation: A Housing Program* (New York, 1952), 222.

65. Some maps are filed with the Cartographic Division of the National Archives. Most of the records, however, are unavailable, and senior FHA officials now deny the existence not only of redlining maps but of any information that might allow an analysis of the spatial distribution of FHA mortgage insurance, other than by country.

66. Quoted in Straus, *Two-Thirds of a Nation,* 221.

67. David Levine, *Internal Combustion: The Races in Detroit, 1915-1926* (Westport, CT, 1976).

68. Charles Abrams, *Forbidden Neighbors: A Study of Prejudice in Housing* (New York, 1955), 229-230.

69. Because the per-capita data are based on 1960 population, the table actually understates the county advantage because it was growing rapidly during the decade while the city was losing population.

70. Over the nation as a whole about 40 percent of FHA mortgages were issued for new homes and 60 percent for existing dwellings between 1934 and 1972.

71. The city was more needy in terms both of the age and the number of its existing structures.

72. Section 223(e) of the National Housing Act (as part of the Housing and Urban Development Act of 1968) gave legislative sanction to relaxing FHA standards in order to permit mortgage insurance for housing in blighted areas of central cities.

73. Essex County was not included in Tables 4.3 and 4.4 because the county contains several affluent suburbs as well as Newark, and there is no way to disaggregate the information.

74. Although FHA mortgage operations were to increase by twenty times in Hudson County, New Jersey, between 1960 and 1976, even at the later date the county was receiving far less than the national average.

75. Livingston, on the western edge of Essex County in attractive and rolling country, was a sparsely settled area of 13 square miles without either industry or railroads. Irvington was adjacent to Newark and was about 80 percent developed by 1938.

76. The original legislation creating FHA authorized an 80 percent mortgage guarantee for homes costing as much as $20,000.

77. FHA, *Washington, D.C. Housing Market Analysis* (Washington: FHA Division of Economics and Statistics, July 1939), 49.

78. Moffett was a former vice-president of the Standard Oil Company.

79. On June 27, 1972, FHA announced that it had "blacklisted" six areas in Brooklyn and had prohibited any additional FHA-insured loans in them. The problem continues. New York *Times,* June 28, 1972.

80. Jonathan Lang, "Problems Facing Urban Renewal in the Fringe City: A Study of Redevelopment Programs in Camden, New Jersey" (unpublished seminar paper, Columbia University, 1972).

81. Mel Scott, *American City Planning Since 1890* (Berkeley, 1968), 401.

82. Abrams, *Forbidden Neighbors,* 234-235.

83. Not only did FHA help move mortgage funds from the cities to the suburbs, but two other housing innovations of the federal government, the Federal National Mortgage Association (also known as Fannie Mae) and the Government National Mortgage Association (also known as Ginnie Mae) have helped more savings funds out of the cities of the Northeast and Middle West and toward those of the South and West. Fannie Mae essentially created a standardized mortgage instrument that all states recognize, and on which banks and other institutions can lend. "Mortgage funds can now move freely across the country to where needed," according to official doctrine. A typical result is that savings banks in the Bronx invest only about 10 percent of their funds in the borough and only about 30 percent in New York State. The rest goes for investments elsewhere in the country, a result that would not be possible except for Fannie Mae. The literature on this is enormous; an excellent place to begin is the special issue on redlining of *Empire State Report* 1 (March/April 1978), 5-33.

84. Bradford, "Financing Home Ownership," 314.

85. Martin Nolan, "A Belated Effort to Save Our Cities," *The Reporter* 37 (December 28, 1967), 17-20. See also Joseph P. Fried, *Housing Crisis U.S.A.* (New York, 1971); and "Ins and Outs of Home Loans," *Changing Times* 13 (August 1959), 26-28.

86. Paul Douglas et al., *Building the American City: Report of the National Commission on Urban Problems to the Congress and to the President of the United States* (Washington: Government Printing Office, 1968), 100-101.

87. Jane Jacobs, *The Death and Life of Great American Cities* (New York, 1961), 301.

88. "Kennedy's Housing Order: Where It Applies, What It Means," *U.S. News and World Report* 53 (December 3, 1962), 68.

89. The standard work on Veiller and his supporters is Roy Lubove, *The Progressives and the Slums: Tenement House Reform in New York City, 1890-1917* (Pittsburgh, 1963). See also James Ford el al., *Slums and Housing, I* (Cambridge, MA, 1936); Robert W. DeForest and Lawrence Veiller, eds., *The Tenement House Problem,* 2 vols. (New York, 1903); and Lawrence M. Freidman, *Government and Slum Housing: A Century of Frustration* (Skokie, IL, 1968).

90. Eugenie Ladner Birch, "Edith Elmer Wood and the Genesis of Liberal Housing Thought" (Ph.D. dissertation, Columbia University, 1976), especially Chapters 1-4.

91. Although designed to provide homes for both urban and rural residents, the North Dakota program was tiny and was operative only between 1919 and 1923. Friedman, *Government and Slum Housing,* 97-98.
Government and Slum Housing, 97-98.

92. 48 Statute 195 (1933), *National Industrial Recovery Act,* Title II, Section 202. See also Federal Emergency Administration of Public Works, *Urban Housing: The Story of the PWA Housing Division, 1933-1946,* Bulletin No. 2 (Washington: Government Printing Office, 1937), 14-16.

93. An excellent discussion of the political trade-offs in federal housing policies can be found in Harold Wolman, *Politics of Federal Housing* (New York, 1971). See also Timothy L. McDonnell, *The Wagner Housing Act* (Chicago, 1957).

94. Actually, American cities could have provided many such instances in the nineteenth century. In New York City, for example, it was common practice to tear down older homes, which already had been divided into as many as a half dozen

apartments, and to build taller and deeper tenement structures in their place. Dumbbell tenements, so called because of the narrow airshaft in the center of the building, represented new private construction in fully developed lower Manhattan neighborhoods. Such housing was obviously not what Ickes had in mind, however.

95. One such project, the 284-unit Carl Mackley homes in Philadelphia, is discussed by John F. Bauman in "Safe and Sanitary Without the Costly Frills: The Evolution of Public Housing in Philadelphia," *The Pennsylvania Magazine of History and Biography* 101 (January 1977), 114-128.

96. The first public housing project in the United States—federal, state, or local—was First Houses, a group of 120 apartments in a series of four- and five-story walk-up buildings at Third Street and Avenue A on the Lower East Side of Manhattan. The 1935 project did not involve federal funds, which were just becoming available. The nation's first federally funded public housing development was a seven-building complex on the Harlem River Drive at 151st Street, also in Manhattan. Work began on the Harlem River Houses in the summer of 1936, and the project was dedicated by Mayor Fiorello LaGuardia on June 16, 1937.

97. The units were in forty-nine separate projects costing a total of $129 million.

98. It was the federal use of eminent domain for housing, not the construction of housing, that was found unconstitutional. *United States v. Certain Lands in the City of Louisville,* 9 F. Supp. 137 (W.D., 1935). See also William Ebenstein, *The Law of Public Housing* (Madison, WI, 1940), 39.

99. Judge Dawson's decision had earlier been upheld by the Court of Appeals for the Sixth Circuit.

100. "Federal Activities in the Housing Field," *Congressional Digest,* April 1936, 104. See also Robert K. Brown, *The Development of the Public Housing Program in the United States* (Atlanta: Bureau of Business and Economic Research, 1960); and Leonard Freedman, *Public Housing: The Politics of Poverty* (New York, 1969).

101. The bill died because F.D.R. did not put pressure on the conservative chairman of the committee, Representative Steagall of Alabama, who would have supported it had the President so requested. The most likely reason for this is that Roosevelt preferred to avoid the political risk of endorsing the bill in the 1936 election year. He was reasonably sure that the portion of the electorate committed to the New Deal would not vote against him on this issue alone, and he did not wish to alienate powerful business interests by openly favoring the growth of the public sector at the expense of the private. McDonnell, *The Wagner Housing Act,* 210.

102. Franklin D. Roosevelt, "A Changed Moral Climate in America," *Vital Speeches* 3, February 1, 1937.

103. New York *Times,* September 3, 1937.

104. For loans of up to sixty years, which meant lower rentals.

105. Housing Files, Franklin D. Roosevelt Library, Hyde Park, New York.

106. *Annual Report of the United States Housing Authority for the Fiscal Year 1938* (Washington: USHA, 1939), vii and 38.

107. The 1937 Housing Act was important in terms of jobs because the unemployment rate among construction workers averaged about 55 percent at the time. Moreover, housing construction is a key factor in economic recoveries because it uses large amounts of capital, labor, and materials. Robert M. Fisher, *Twenty Years of Public Housing: Economic Aspects of the Federal Program* (New York, 1959), 229.

108. In Britain, where the Housing Act of 1919 started public housing, the municipality itself is the "housing authority." And in Japan, the national government buys inexpensive land in very distant areas as the only practical means of acquiring space for public housing projects. Smith, *Housing,* 127.

109. Between 1977 and 1979, Secretary Patricia Roberts Harris of the Department of Housing and Urban Development attempted to use financial incentives to encourage suburbs to accept a fair share of public housing.

110. Bauman, "Safe and Sanitary," 116-125.

111. Quoted in Fisher, *Twenty Years of Public Housing,* 96.

112. *Congressional Record,* April 21, 1949, 4840.

113. *Congressional Record,* April 21, 1949, 4852.

114. Between 1969 and 1974, President Richard M. Nixon frequently affirmed that he would not use the financial leverage of the federal government to compel suburbs to accept low-income housing against their wishes. And in 1971, the U.S. Supreme Court upheld the constitutionality of state laws requiring approval in a public referendum before low-income, subsidized housing could be built in a community.

115. When a civil-rights-minded mayor, such as New York's John Lindsay, attempted to force a public housing project on a middle-class area, as in the Forest Hills section of Queens, the result was simply an accelerated white flight to the suburbs. Thomas M. Gray, "Daley News: Chicago's Public Housing Fiasco," *The New Republic* 164 (April 3, 1971), 17. See also New York *Times,* October 1, 1973 and April 16, 1976.

116. For example, in 1869, the New York State Assembly passed a law permitting the city to pay half the cost of street openings.

117. Although the 1901 law was technically retroactive, as a matter of practical fact it did not apply to "old law" tenements.

118. Clawson, *Suburban Land Conversion,* 41. The Housing Act of 1964 (Section 121) authorized FHA to pay the owner of an FHA home any costs incurred in correcting "substantial defects" in the home.

119. The most important federal inducements to dispersal have related to transportation and taxation (mortgage and tax deductions) policies. In those cases, as in the case of housing, the real beneficiaries have been financial institutions and large businesses rather than the presumed beneficiaries, homeowners or the poor.

120. In metropolitan area size, St. Louis ranks in the top dozen, but even there the region has lost population in comparison with most other areas.

121. There were signs in the late 1970s that St. Louis would become part of the nationwide trend toward urban restoration. Houses that only a few years ago were virtually worthless have once again been occupied.

5

SOME ELEMENTS OF THE HOUSING REFORM DEBATE IN NINETEENTH-CENTURY EUROPE

Or, On the Making of a New Paradigm of Social Control

LUTZ NIETHAMMER

IN SEARCH OF THE MEANING OF HOUSING REFORM

Most histories of the evolution of social policy in Europe share three characteristics: First, they tell a success story, the rise of the modern welfare state, even though its development may haven been slow and protracted. Second, they relate the concepts by which social reform was initiated by political groups, organized interests, and schools of thought, motivated initially by charitable or humane concern for the victims of the industrial revolution, and later concerned with reconciling industrial relations and income inequities. Third, if housing reform is discussed, it is treated only superficially.[1] This comes as a surprise for two reasons: Housing was the most expensive item for consumers, constantly in too short supply to meet the needs of the poorer strata in urban society during the first century of social policy development. Also, the housing question was debated more frequently and at greater length than any other social policy question. The three aspects mentioned seem to be related, for housing reform, at least up to

World War I, was hardly a success story. Nor is the history of housing reform fully to be explained by the dominant interests of capital and labor or by the programs of political parties or charitable institutions.

The historiography of nineteenth-century housing in Europe complements the broader issue of social reform, though not in all its aspects. Most of the studies, even those informative works that cover a longer span of time,[2] approach the subject matter from a rather narrow, professional point of view—like those of architecture, city planning, public health[3]—or concentrate on certain compensatory strategies to meet housing needs, like paternalism, philanthropy, cooperative initiatives or state intervention.[4] They take for granted the importance of environment and reproduction in a society dominated by industrial capitalism, but this becomes evident only in retrospect. Mostly they lack a broader perspective on the social uses of the habitat, on the drive to transform housing into a political problem; or they fail to place the housing debate in a wider framework of social change and control. Though it is clear that the early reform legislation was hardly enforced and on the whole utterly ineffective, and that prior to World War I housing for the urban poor lagged behind urban growth and rising living standards, historians have tended to compensate for this by concentrating on the ever more complex and sophisticated debates on the problem, as well as on the legislation passed to prove progress, if only on paper, by citing single examples that had little or no impact on the overall situation.[5]

This plethora of discussions without social relevance and this sense of progress without qualification, together with the contradictions between suggestions and results, require I think, a new look at the debate, in an attempt to reduce its complexities, to rediscover its dynamics, to identify its contributors, to define its framework, to analyze its relation of theory to practice, and to speculate on its significance. I propose to look at the housing-reform debate as an attempt to moderate the class struggle, introducing, beyond welfare economics and the institutionalization of industrial relations, physical dimensions: public health, social space, and the family as reproductive transcendency.[6] The housing debate, then, may be conceptualized as the experimental formulation of a new paradigm of social control,[7] combining culture and nature and offsetting the predominant mixture of politics and economics. The resulting program, however, was in general too expensive to be acceptable to the dominant private interests of rising capitalism and went through a series of dismantling experiments to make it less costly, until the rising lower-middle and upper working classes advocated it in their demand for social space, and political intervention effectuated it in a much modified form.

The present essay discusses the debate on housing reform as the lengthy evolution of a new paradigm of social control in three ways. First, it presents a theoretical concept of the basic cultural dialectic of urban society during the rise of industrial capitalism. I hope that such an interpretive tool will be useful to organize widely scattered evidence and to suggest that cultural and economic antagonisms in that society, while not identical, were closely related. Second, it discusses three authors, chosen as analytical examples from a very widespread and differentiated debate in Europe,[8] identifying the implications of their suggestions regarding health, space, and family, and looking for a pattern of thought that generated common perspectives. Last, it places a new paradigm into the socioeconomic process, in search of an agent with complementary interests and sufficient political weight to break the economic deadlock that prevented any real housing reform in nineteenth-century Europe.

At this point I should like to admit that I used, and profited from, the rich survey literature that I criticized above for its lack of analytical structure and meaning. I am especially indebted to some detailed case studies on the dialectics of reform,[9] as well as on such related subjects as the family, sex, the body, the perception of cities or the working class, planning, and social policy in general.[10] And I should like to apologize for the raw and sketchy nature of my attempt at identifying some basic guidelines, which omits or even does violence to a mass of details. This is all the more pertinent since I shall look at housing reform at a European level. However, the international parallels are so striking that an approach limited to the national or local level, while it might do more justice to the individual contribution and its environment, would risk missing or obscuring the fundamental social meaning of that new common paradigm of a physical dimension in social reform, and the socioeconomics, first of its delay and then of its breakthrough.

URBAN GROWTH AND CULTURAL POLARIZATION

Regardless of a more complicated urban stratification, the cultural dialectic of the growing European cities in the nineteenth century can be reduced to two polarizing sociocultural subjects. Certainly, they do not explain all of the social reality, but they are informative about the perceptual framework of sociospatial reform.[11]

After the dissolution of the former household economy[12] in the course of the industrial revolution and the expansion of commerce and of administrative services, middle-class life became increasingly channeled into a system of functional division and localization, tending to the formation of compensatory polarizations. Some examples may be cited: in the city, the zoning of space devoted to public life, to business, to production, or to residence, each with its own architecture and symbolism; in the house, the differentiation of rooms, each with a special function and an appropriate design, including an educative and representative imagery;[13] the distance between the semipublic drawing room and the private spaces used for physical reproduction; the separation of a male empire outside the home, oriented toward the management of production and the accumulation of property in an atmosphere of power, calculation, and performance, and a female laboratory for the management of consumption and reproduction in a climate of intimacy and emotions.

This functional and spatial differentiation reflected the self-alienation of the bourgeois, accompanied by the making of a civilization centered on consumption and, more important, on commodities and other purchasable objects and services. In other words, bourgeois individuality became identified with buying power and with the selection and the arrangement of objects, and provided a civilized ecology conditioning the next generation. Under the skin, however, the bourgeois did not feel at ease. His first personal and sociable experience, the awareness of his own body, was a precarious one, because the body was no longer accepted as part of nature or used as a source of energy, but was substituted by social relations and repressively conditioned in a long, formative period, reducing it to a sensitive, painfully embarrassing, and even dangerous object.[14] The protection, containment, and organization of this precarious physical and sentimental existence, the Achilles' heel of the bourgeoisie, was the most important function of the home. The middle-class notion of the home as one's castle reveals the ambivalent desire for defensive domination.[15] The growing density of urban coexistence and the inevitable contacts with the savage crowd were seen as potential threats, by physical violence, to the social order. Hence, the urban strategy of zoning, the corresponding movement of the immigrants into the ancient city centers, and the middle-class escape to some west-end, suburbia, or country cottage;[16] hence the reestablishment of carefully disciplined relations, for instance, in sports and parks.

The intense and complex feelings of the middle class in this matter, which I shall illustrate, reflected its confrontation with a counterculture perceived as savage, uprooted, unrestrained, lacking property and providence, liquid and filthy, a swamp breeding violence and sickness—all metaphors used to describe a danger considered both physical and social.[17] Has there been parallel to these traumatic projections, a social reality to the urban crowd—liberated from the restrictions imposed by personal bonds in small patriarchal communities to the independence and the mobility of the labor market—young, often single, poor, with an extreme residential instability?[18]

Above all, this was a physical and collective form of existence. Bodies were continuously in motion and in contact—at work, in the street, at home, from childhood to an early old age. Bodies serving as sources of energy, as tools, as capital. There was proximity without privacy. Often, partially naked bodies crowded in factories and dwellings, where the lack of space prohibited functional differentiation or any separation of sexes, ages, even of people from animals; of space for eating, sleeping, working; of the healthy from the bedridden; at times even of the living from the dead. Sexuality here may have been less of a secret explosive than among the repressed and disciplined middle class, because these bodies were less intangible, more exhausted, and sexual intercourse was less concealed and more instinctive. And in this everyday physical routine, leisure was less directed into preformed and sophisticated channels; it had more the aspect of interruptions, erupting not only in feasts or marches, but even in unemployment, crime, strikes, drunkenness, or illness.

Under these conditions, the notions of alienation and civilization or culture take on very different meanings. Deprived of property and thereby of tangible representations of their identity, the vagrant poor and, above them, a large part of the working class did not have a materialized civilization, but a culture of sociability and accommodation in the midst of an alien domination of all the means of production and reproduction. The concept of a culture of poverty,[19] then, as I would like to use it, denotes in essence an immaterial structure of interpersonal relations, performing adjustments in changing collective forms of existence and possessing the ability to endure a world that is designed, controlled, and owned by others. Alienation here does not result in split personalities, but in a deprivation of goods and means, and in the opposition to or evasion of a social and public order reflecting and strengthening this one-side allocation of material wealth and power.

The interdependency of habitat and family among the urban poor is also very different from that of the middle class. Here, it was not the family structure that materialized in a functional division of spatial representations; on the contrary, there were alien spaces, narrow, multifunctional, changing, and exchangeable, to which some group of persons had to conform, be it a family or another group brought together by circumstances of mutual usefulness. The family was only one form among the cooperative units of reproduction, certainly dominant during child rearing, but usually absent during a long youth and quite often in old age. Even proper families were not closed units, but, in case of need, semi-open to lodgers. This situation arose most often when families could not afford the upbringing of their children. Under these conditions, traditional roles in the family could be distributed differently, members could become relatively independent, and intermediary forms could develop between "private" life and class comradeship. In other words, the semi-open family[20] could be a "home" that did not function as a shelter from the risks and strain of work and public life, but was just another space among others, reflecting a unity of collective deprivation and material alienation in all dimensions of life and contributing to the emergence of a culture of evasive accommodation and collective resistance.[21]

THE INTRAMURAL SAVAGE

"The man without property needs a constant effort of his virtue to get interested in an order that does not conserve him anything. A country where the proprietors govern has a social order; but one where the unpropertied govern is in a state of nature."[22] When these words were spoken, in the third year of the French Revolution, they still had the flavor of a philosophical tract, distinguishing nature from civil society and eliminating the unpropertied from the electorate because they had nothing to lose. Yet, they may provide a good starting point, because they very aptly express the thinking that structured bourgeois perceptions of the proletariat. Nature, obviously, is something dangerous and alien to bourgeois society, which relies on economic categories. Poverty, however, is essentially nothing within the economic order, but a concept of a different dimension. It robs a person of civil roles and reduces him or her to a body. Educating people to the exercise of virtue may lead to integration. But without their accession to something they could lose ("the novitiate of property" for the "modern barbarians," as a later French speaker would call it)[23] hoping was not enough.

In the first half of the nineteenth century, this perception prevailed, but the physical encounter with poverty became increasingly painful for the rising bourgeoisie, because the accumulation of its property corresponded to the urban concentration of poverty, bodies of a workforce with no economic power but only physical strength. The medical explorer coming in contact with these intramural savages, in 1834, hardly dared to put his perceptions in terms of his own culture: "The third and last class, that of the proletariat, is of a relatively enormous extension and with few honorable exceptions, [has all the] profound ignorance, the superstitions, the ignoble habits and the lacks of manners of wild children. Its rudeness, its brutality, its lack of providence, its wastefulness at extravagant feasts and orgies, which impair so much its well-being are—and I say this without prejudice—undescribable; the depiction would become too disgusting."[24] Within a short time, the literary public got quite a lot of the indescribable in social novels describing as well as bridging the gulf between the "two nations" that Disraeli had found coexisting in early Victorian England: the rich and the poor.[25] And the public obviously liked the thrill that these descriptions of the other nation provided, all the more since the novelists saw the worker and the poor, up to the naturalist novel of the later nineteenth century,[26] from the bourgeois perspective as physical, pre-civilized, or wretched creatures, and their accounts could be read in an armchair at a safe distance.

The environment of these creatures was a similarly shocking counterworld to the rising differentiation and functional discipline of bourgeois space, a hotbed of social dangers and, at least in our topography, the fault of the poor. "The shelter [of the proletariat] consists either of old, hermetically closed decaying masonry or of barns, open to all the elements, filthy, dilapidated, and narrow, or they are packed together, living at random, breathing stinking gases (veritable winds of poison), swarming and growing exuberantly in a shameful void, violating...in their cynicism all shame and delivering finally thousands of victims of their debauchery and corruption to general impoverishment or public hospitals."[27]

As such, this view of the crowd and the urban poor as immoral animals that had to be contained is certainly not new. What is new are the dimensions of the problem: the dimensions of numbers at a time of massive rural influx into the growing cities, like the 1840s, when the marginal threatened to become a majority; the dimensions of spatial constriction and vanishing physical distances, with the poor swarming, menacingly omnipresent, in narrow streets and dark back-buildings, in cellars and attics, using every rotten barn and broken-down hut, especially on the outskirts of the attractive metropolitan centers. The urban bourgeois already felt that he was sinking into a foul and rising tide, when again new dimensions added imminent physical threats: the spread of epidemics, especially cholera, and the rising political unrest and violence among the urban masses, which culminated on the Continent in 1848. The physical threat of the underworld could no longer be ignored as a state of nature on the margins of civilization. On the contrary, it had become part of a growing, instead of a fading, feature of the urban culture.[28] Similarly, as in their relations with the indigenous savages of other continents,[29] the middle class answered the challenge of the intramural savagery with a combination of repression and acculturation, often translated into medical terms. Repression, however, was much the cheaper expedient in the short run, and therefore the problem receded for long periods or manifested itself periodically, despite the repression of riots and the political organization, public health agitation and the progressive excision of slums, the widening of streets, the introduction of railways, and the redevelopment of city centers together with the planning of massive expansion schemes, sanitary provisions, and social distance.[30]

The threat of the savage from within was not overcome by urban growth, nor did that basic structure of perceptions which transformed economic disadvantage into threatening bestiality fade away. On the contrary, the density and complexity of such experiences reinforced and intensified the basic approach, not only to amplify its grasp of the problem. I can give here but one example for the persistence of this approach until the end of the century (and, really, beyond), which I take from the German Ebenezer Howard, Theodor Fritsch, who advocated the garden city in 1896. The basic assumptions of bourgeois liberalism, however, are now longer with us, because they stand accused of breeding savagery, but Fritsch's call for authoritarian order could even be found among the more liberal contributors to our debate. "Isn't an environment that displays in all its phenomena irregularity and opposition to order, and ignores all systematic planning, bound to breed in those men growing up in it the spirit of unreason, confusion and a lack of discipline? And would not, on the other hand, a city, the result of a clear and foresighted spirit,...constructed in noble regularity and beauty, have edifying and settling effects on the human spirit? In wilderness and chaos the wildest and rudest instincts are aroused, while even the wild beast loses much of its unruly attitude when feeling the barriers of a superior settling power. The sense of order, the power of harmony have taming effects even on the rudest of minds."[31]

Most responsible for the stagnation of spatial and hygienic problems in the middle of massive urban change was the almost constant immigration and the persistence of other forms of mobility, which added still another dimension to the challenge facing bourgeois civilization. The importance of mobility for the nineteenth-century city need not be proven here, since several historians have shown the extreme fluidity of the living habits of a great part of the lower urban population.[32] But instability was not only transitory and affected not only the poor. Rising land values in the growing cities often led to a constant escalation of rents, so that even the lower-middle class could no longer afford them and had to move. In 1857 a writer on the housing shortage in Vienna observed, "During recent years the feeling of stability has been completely lost by the Viennese. No suburban inhabitant feels secure, from one year to the next, on his 'soil,' in his street, within his walls. And the inhabitant of the inner city no longer knows, at what distance and in which hitherto unknown region of the widespread metropolis he will, in a half year's time, rest his tired body. Therefore [homes] can no longer be spoken of...dwellings, at best.... [They are] temporary shelters,...tents of stone constantly changing their inhabitants. Nobody can boast about or rejoice in a 'home'; nobody can dare to choose or establish his house to provide for future events in his family. So an undercurrent of unrest and anxiety has affected the whole population of Vienna. The transiency of...nomads has replaced the calmness of a settled bourgeois existence, and the metropolis of Austria lodges within its walls a population constantly on the move, violently crowding and striking against each other, discontented with the insecurity of its domestic life and the high rents of their transitory lodgings; a population that increasingly loses any sentimental attachment to its native soil, its [sense of] participation in...local and public matters, and its sense for order and peace."[33]

What is interesting to me in texts like this—and there are many, many more in the literature on housing and urban reform—is the complexity of their perceptions and notions. Not just housing scarcity or human mobility are being dramatized to arouse public concern; rather, the sensual perception of a whole culture being endangered implies a set of private and civic values, and focuses its anxieties on the precariousness of the body, of social space, and of the socializing institutions that link biological reproduction with the general public order.

MEDICAL TOPOLOGY AND POLITICAL ECONOMY

Edwin Chadwick[34] had participated prominently in the struggle to repress the mobile culture of poverty by limiting public relief. Beyond fact-finding and agitation, his contribution to the new Poor Law made him an early predecessor of this approach to social control and political economy. The workhouse combined institutional control of the hard-core poor with an incentive to the rest to escape to the labor market and to submit to industrial discipline, thereby supporting economic growth and cutting public expenditure.[35] At least in the short run, however, this hardly led to the integration of the poor into civil society, but provided a supply of abundant, cheap labor clinging together in urban slums, the breeding ground of contagious disease and the source of improvidence and political unrest. It may be open to discussion whether Chadwick's subsequent interest in the problems of public health was due to his logical and empirical mind or was inspired by his own experience with an almost fatal encounter with infectious epidemic fever. What is interesting, however, about his "sanitary idea," is again his approach to the problem of marginality, his perception of health in spatial and economic terms, and his translation of the human challenge of high death rates into the language of taxpayers' democracy.

All over Europe medical topology had already begun, by the analysis of more or less crude correlations, to link the massive outbreaks of diseases and short life expectancy to specific local conditions.[36] Identifying slum and working-class quarters as the enclaves of early death may have inspired charity, but it could also deter social action, since it could be argued that nature eliminated those who remained in a state of nature and did not acquire the standards of civilization. Chadwick incorporated the survey techniques and their findings into the more general framework of the health and wealth of nations, often with problems not restricted to the poor, but effecting fatal consequences for society at large. This argument is well known—as applied, for instance, to contagious disease—and needs no amplification here.[37]

Interested in the more complex and dynamic ideas of the sociospatial reformers, I have identified a different line of thinking from among the wealth of evidence and argument presented in Chadwick's famous 1842 report on the sanitary condition of the laboring population.[38] "The depressing effect of adverse sanitary circumstances on the labouring strength of the population and on its duration, is to be viewed with the greatest concern, as it is a depressing effect on...the chief strength of the nation—the bodily strength of the individuals of the labouring class. The greater portion of the wealth of the nation is derived from the labour obtained from the application of this strength." With these words, Chadwick summed up his analysis of the "comparative chances of life in different classes," and he went on to show that British industrial advance was at stake when the physical health of labor was deteriorating under the influences of crowding, promiscuity, and lack of air and running water. Pointing to an elite group of English workers employed in the construction of railways in Germany, who outperformed their German colleagues in productivity and wages, he stated, "Skill and personal strength are combined in an unusually high degree in this class of workmen, but the most eminent employers of labour agree that it is strength of body, combined with strength of will, that gives steadiness and value to the artisan and common English labourer."[39]

So, next to the physical condition of the labor force, it was industrial morale—that is, maintaining social peace and work discipline—that was decisive in the international economic competition, and again he saw this related to the body and its well-being. To him, class struggle did not arise out of exploitation, and even seditious agitation was not the prime cause of political unrest; rather, it arose from lack of experience. Discussing the repression of trade union agitation in the district of Manchester, he quoted the employers of that city as saying that among the working population only "those of mature age and experience...were intelligent, and perceived that capital, and large capital, was not the means of their depression, but of their steady and abundant support. They were generally described as being above the influence of the anarchical fallacies which appeared to sway those wild and really dangerous assemblages." This core of industrial peace, however, was in short supply and could not control the majority of their younger comrades. "The proportion of men of strength and of mature age for such service (as special constables) [was] but as a small group against a large crowd, and that for any social influence they were equally weak." Labor just did not live long enough to appreciate the beneficial effects of capital, because of the "disappearance by premature deaths of the heads of families and older workmen." "The predominance of a young and violent majority was general," especially in London, where the "mobs" were generally between 16 and 25 years of age and, when coming from the East End, were "proportionately conspicuous for a deficiency of bodily strength, without, however, being from that cause proportionately the less dangerously mischievous."[40]

Thus, if sanitary reform could restructure the slum areas, making them a more healthy environment, it offered two long-range advantages: It would improve the working person's bodily health and add to his economic value to sustain industrial growth, and it would provide for an ordering of proletarian age groups, to bring the inexperienced and violent young again under the moderating restraint of their elders. In short, Chadwick stressed "the importance of the moral and political considerations, viz. that the noxious physical agencies depress the health and bodily condition of the population, and act as obstacles to education and to moral culture; that in abridging the duration of the adult life of the working classes, they check the growth of productive skill and abridge the amount of social experience and steady moral habits in the community; that they substitute, for a population that accumulates and preserves instruction and is steadily progressive, a population that is young, inexperienced, ignorant, credulous, irritable, passionate, and dangerous, having a perpetual tendency to moral as well as physical deterioration."[41]

It is clear that Chadwick did not turn to public health primarily because he cared about the sick, but because physical deterioration and social unrest were dysfunctional to a system centered on two constants: the progress of science and the progress of capital as the promoter of the market economy, of growth, and of welfare. The basic principles of capitalism were kept inviolate; if one were to cure its ill effects, one had to open up new dimensions. Almost anything around these fixed stars could be reformed—reoriented to fit into their dynamic system—including administrative structure, city landscapes, and the moral or physical reproduction of the population at large. Environments had to be washed and ventilated and cleansed of filth and waste; stagnation had to be replaced by circulation; moisture and dampness had to be drained and eliminated; resisting culture had to be dislocated or controlled. Chadwick showed an almost priestlike fanaticism against everything stagnant and not conforming to his abstract fetishes. He was neither a capitalist nor a scholar; he was not interested in private property and his use of statistics was as unscientific as his dogmatic application of medical theories. Certainly, his personality would make an interesting psychological study, but such is not the concern of this chapter. What I should like to point out, however, is that in Chadwick's post-Poor Law views on political economy, the essential complement to capital was not just the availability of labor, but the conservation and improvement of the bodily condition and morale of the poor, and that meant combining environmental control with a systems approach: public health.[42]

Through the second half of his life, Chadwick sought acceptance of his reform proposals as capitalism (as an abstract principle, certainly not as the interests of specific individuals). His calculations of the social savings to be gained through investing in sanitary reforms in order to lower the death rate may at first seem to be the attempt of a humanitarian to sell his proposals to a frugal middle class. But a study of his later writings, when he had become an old and frustrated man who kept making ever-new proposals and rigidly dogmatic statements, provides ample evidence of the fixed, hard core of his thinking. Because they are too long, I shall refrain from quoting from his tracts on "Health versus War" (with subsections on "Military Gains from Sanitary Science"), "Savings by Sanitation in One City," "Utilization of Roof Space" (for the collection of soft water and the recreation of the head of the family "in the oriental fashion"). "Circulation or Stagnation," "Results of Sanitation in India,"[43] or "Sanitation as a Remedy for Irish Discontents" (where, for instance, he suggests a correlation between the extent of political protest in certain counties with the percentage of one-room huts, mud hovels, low age averages in the population, a high crime rate, and a lack of drainage in the fields, and proposes santiation as the most important measure against the "naturally resulting disturbances."[44] I shall give but one example of how he transforms death rates into profit rates in a tract on "Local Self-Government in Sanitation": "Take the common case of a slum with a death-rate of 40 in a 1000, and of the expenses it entails on the community. There will be 25 funerals at 5 £ each, and at least 20 times that number of cases of sickness at 1 £ each, every 5th case on the average being that of an adult, entailing on an average, a loss from disability for 2½ weeks at 1 £ per week, making 1875 £ p.a. of expense for 1000." Then he calculates that, by better sanitation, the death rate could be lowered by 10 in 1000, and transfers his results to a population of the size of Manchester. The resultant savings would equal the annual 5 percent interest on more than £1 million. He goes on to compare these savings with the investments needed for sanitation, calculated per house, and strikes an even balance, "which represents, at a very moderate estimate, in one factor only the loss due to their absence, namely the sickness, the loss of labour, and the funerals, to say nothing of the cost of the citizens, not to speak of the misery and pain." This really comes last, after two pages of calculations, and, in fact, he does not speak of it. However, in the next paragraph he converts the "cost of the citizens" into pounds, assuming, under English conditions, a life lost to equal a loss of not less than £100; he does not forget to mention in passing that in America they even estimate it at some £200. Then he produces an aggregate sum of £148,000 per annum lost by those dying from lack of sanitation, and adds "an excessive expense of police and penal administration, which is chiefly occupied with heavily death-rated populations and places"—only to blame in the end for all these capitalized losses "our local self-goverment..., the worst of any we have."[45]

THE SOCIAL USES OF MULTICLASS SPACE

"The habitation provides in most and the most important respects the conditions of the weal and woe of our existence. It provides us with, or deprives us from, irrecoverable sources of life—air and light. Its quality conditions our well-being so continuously and with so lasting effect that it is above all habitation, on which the caring eye of public health care should be directed."[46] This rather general statement sets forth the basic assumptions of sanitary housing reform, failing only to mention water circulation as a basic need. According to it, housing conditions are most important, that is, more important than working conditions. Essential here is the reproduction of physical life: the instrument of reform is the reintroduction of natural resources into urban life; and the agent of reform is neither the individual nor a group, but an entity placed above with public powers and parental feelings (note its "caring eye"). This higher authority, which need not be but is, in this case, the state, has, however, no economic powers to provide housing for the homeless and overcrowded. It works by inspection, persuasion, and regulation, so that air and light, elements free of cost and in general abundance[47] though increasingly rare in urban life, may flow into "our" existence. It puts restraints on the economy in the name of ecology, authorized by the alleged pre- or metasocial interest of everyone's natural existence. In fact, the sanitary reformer rearranges the relation between two invariables: the physical and the ecological, endangered by a stratified society and the capitalist economy, especially in its urban setting. For this rearrangement the reformer needs additional public powers, in order to introduce ecological lubrication into the urban machine of bourgeois society, to offset its sickening or even death-dealing results.

The quotation cited above reflects the thinking of almost any sanitary housing reformer, advocating public health inspection, slum clearance, building regulation, and other measures as weapons against high death rates, urban epidemics, physical unfitness for work and military service, alcoholism, violence, and criminality. In the long run, this line of thinking is aimed at a low population density and family housing in a garden-city setting full of light, air, and sewers.[48] But the above quotation comes from James Hobrecht, who in 1868 defended the system of the *Mietskaserne*, the multistory, high-density tenements. He is best known for drafting a major expansion scheme for the city of Berlin in 1862,[49] whose plan for big boulevards with deep land plots, allowing for highrise court housing, encouraged capitalist land speculation and building and provided the major model for a new high-density use of urban space on the Continent, along with the Hausmannization of Paris.[50] Defending the *Mietskaserne* and sanitary reform at the same time presented no contradiction to him. Obviously, few, but wide, streets (paved, airy, and with sewers) complemented very densely populated housing complexes. But he also thought that the *Mietskaserne* was a step toward sanitary improvement not only in that it made "running water" economically feasible, but also by creating a constructive social environment.

Hobrecht compared his model with the English, since small houses were the ideal for almost all bourgeois housing reformers in Europe, and admitted that the one-family house produced the best defense against bad moral and physical influences, but that it could only be realized for the middle class, and, in fact, proved to be advantageous only for the rich. In reality, this ideal went along with residential differentiation by class and the division of the society into what Disraeli had described as two nations. Only the police and some depraved poets, said Hobrecht, would dare to intrude into an English working-class neighborhood, because vice and crime everywhere accompanied poverty. The middle class feared any contact with the poor and tried to assuage its conscience by financing institutionalized philanthropy. The condition of the poor, the most numerous class in urban society, however, had to be the major consideration in urban housing reform. Because of their ignorance and their lack of self-help, the poor suffered the most from bad housing conditions, and in turn affected the more prosperous classes physically and mentally. "Not separation, but penetration seems to be, from ethical and therefore from state consideration, should be, the leading principle." What could the *Mietskaserne* do about it?

Hobrecht describes a typical *Mietskaserne,* where a flat on the first floor rented for 500 thalers a year; two flats on the second and on the ground floors rented for 200 thalers each; two flats on the third floor for 150 thalers; three flats on the fourth for 100 thalers; and various other flats in the cellar, the attic, the courtyard, or the back-buildings for some 50 thalers each. "The sight of and contact with...all gradations of poverty and destitution serve the rich and prosperous as moral education, while separation would lead either to callousness or, in the case of more sensitive natures, should this contact occur one day, to a false and nervous humaneness." Middle-class charity should be practiced not though impersonal contributions and institutions, "but...in person-to-person contact and should consist in advice, admonition, and support suited to the special nature of the need." Hobrecht went on to illustrate how the multiclass tenement worked as an integrator of functional and educational interdependences:

In the *Mietskaserne,* the children living in the cellar walk through the same entrance hall to their "Freeschool" as those of the higher official or the merchant to their Public School. Cobbler Wilhelm from the garret and old bedridden Mrs. Schulz from the court-building, whose daughter earns a scant living by cleaning and needle work, become known on the first floor. In one instance a cup of soup helps the recovery of strength during sickness; in another, an item of clothing, or effective assistance to get a free education [is given], and all this help results from the comfortable relations among the inhabitants of the same building, even though their status may vary in the extreme, and [it has]an ennobling influence on the donor. And in this extreme social range [are] the poorer elements from the third and fourth floors, social classes of the utmost importance for our cultural life: the government official, the artist, the scholar, the teacher, and others [in whom]...can be found the intellectual elite of our nation. Forced to work continuously and to make frequent sacrifices [and doing so]...in order not to lose and possibly to improve the place for which they struggled in society, they cannot be overrated as teaching elements in setting a good example, and they have an encouraging, stimulating, and therefore socially useful infuence on those who live next to and mix with them, be it...only through their mere existence and their silent example. The more prosperous, on the other hand, because of their cleanliness and their social graces, not to mention their better manners, the result of their more careful education...exert a moral influence on the poor and less advantaged. In an English working class quarter a mother may leave her children unwashed, uncombed and ragged; but the mother from a cellar dwelling in a *Mietskaserne* would be frightened to do so because she knows herself to be watched and exposed to criticism from her betters.[51]

Hobrecht ends by pointing to a variety of odd jobs that the more prosperous can distribute among their poorer neighbors to improve their income and to let them use their skills.[52]

This passage may be remarkable or even unique in defending the social value of the *Mietskaserne*, singled out in almost all the rest of housing reform literature as a symbol of the social and ecological vices of urban density, exploitation, and destitution.[53] Yet, its way of thinking, its methodological approach, and most of its value judgments are widespread in this literature.[54] A few features should be stressed once more.

Even the sanitary reforms of the early days rarely were advocated for their own sake. Usually, they were part of far-reaching aspirations for social reform of a very special kind. The main problem was poverty (rather than workers)—more precisely, poverty in the framework of urban growth and population density close enough to become a focus of middle-class fear (rural poverty does not arouse the same concern), but at the same time distant and massive enough to have escaped from earlier forms of social and cultural control, becoming independent, strangely like the native population of the colonies, and governed similarly by sheer force and apartheid. Contrary to the socialist, who tries to use this situation to build an alternative society or even tries to share its experience, our reformers look at it from the outside and consider poverty not as an alternative culture, which is exactly what they want to prevent, but as a state of deficiency that should be broken down into manageable units to be reintegrated with the respectable classes along general, presocial, natural structures like space, health, and kin rather than class, wealth, or power.

Reform, then, is centered on the private life and aims to appease class tensions by ignoring their basis in the sphere of production and by experimenting with mechanisms of social control—ranging from separation to integration of classes in the sphere of reproduction—broken down into small entities like families, neighborhoods, tenement populations, and the like. Most of the early sanitary reforms practiced segregation, especially by destroying the inner-city slum areas and relocating their population in single-class residential suburbs. Especially in the metropolitan areas with their constant influx of the rural poor, this repressive attempt was rather unsuccessful, because inner-city slum areas survived through the 1870s in an even more compressed form, and lower-class suburbs reproduced within a short time much of the culture of poverty and of other segments in the working population, now controlling, unchallenged, wider urban districts.

Yet, Hobrecht argues already for a postliberal conservativism, that is, for paternalistic integration based on a realistic appreciation of the short-term and complex character of the various classes' needs and their functional and moral interdependence within the given structure. Again, it should be noted that this sort of approach, though it may have been influenced by a conservative, utopian view about the peaceful working of a preindustrial, multiclass household,[55] did not derive from ignorant traditionalism. Hobrecht knows England, and his irionic remarks on segregation echo a whole chorus of English critics, often people with the most intimate knowledge of the conditions of the poor. Hobrecht stands for the conservative reaction to the urban and social results of liberal capitalism; obviously, he also stands for the new lower-middle class, as the nation-state's real defender of morality, service, and efficiency, fighting for and defending its "social space."

But this sort of conservative realism has its limits: It aims to reinstate the complexities of life against the economania of liberal practice and socialist theory, but it lacks a grasp of the economic magnitude of its projects. Before state intervention, most of the housing reform proposals never materialized, or produced only models with marginal or no influence on the building market, because they were formulated in the interest of the society at large, as it was perceived by the respective planner. This interest, however lacked economic support and proved to be unable to fight the private interest of the property owners in control of the economy.[56] This was largely true even of the radicals like Chadwick. In Hobrecht's case, his vision of the multiclass *Mietskaserne* was defeated by the upper classes in Germany, as everywhere else, which moved multiclass environments for some west-end or, later on, suburbia, reducing the *Mietskaserne* to a far less harmonious meeting place for the various strata of the intermediate and lower classes only, with strong tendencies at demarcation among them.

A NOTE ON SOCIOSPATIAL REFORM AS AN INTERNATIONAL DISCOURSE OF OUTSIDERS

That Hobrecht had drafted his city expansion scheme for Berlin with Haussmann's plan for Paris in mind—that he discussed residential differences in England and followed Chadwick in his call for urban air and water circulation and for

a centralized government department of public health—was
certainly not exceptional. Bourgeois social reform—especially
the debate on sociospatial strategies—was an international
movement with a constant exchange of experiences, through
books, articles,[57] and visits, and even special organizations. Its
international references are indeed so frequent that they may
come as a surprise to those who associate the nineteenth-
century bourgeoisie with nationalism and are familiar with its
fierce denunciation of the timid and rather unsuccessful in-
ternationalism of the left. This exchange was no one-way
import into backward countries, but worked in all directions,
often legitimizing its position by pointing to an allegedly
successful example abroad, exploiting national rivalries in the
interest of social reform. The explanation for this phenomenon
is to be found in the position of the reformers themselves within
their own environment, where they were usually excluded from
the ruling class and wealthy establishment, though some of
them became quite prominent. Against the dominant class
interests they were voices in the wilderness, suggesting
strategies against social dangers, often personally experienced;
a comfortable middle class tended to ignore them as interference with their short-term interests, and only became
alarmed by intermittent waves of epidemics and revolutionary
upheavals. On the other hand, the reformers certainly did not
share the working class's leftist sympathies, which they most
often characterized as symptoms of a sick social body.

Their cure for economic class struggle was to introduce a new
paradigm of environmental and structural control, stressing the
national, spatial, and educational conditions of reproduction
instead of the relations to the means of production, social
biology as an essential supplement to political economy, social
microorganization instead of political decision or the formation of organizations, and they tried to legitimize their
position by technical expertise and empirical research.

THE TRANSCENDENCY[58] OF REPRODUCTIVE UNITS

The French engineer, pioneer of empirical social research,
and conservative reformer, Frederick Le Play,[59] best represents

all these features and played an influential role in this growing debate on reproductive acculturation. Though he contributed only occasionally to the housing debate proper, his methodological achievement in imposing conservative reform derived from empirical scholarship on antagonistic class positions, and his theory of the family as the agent of social reconstruction made him, regardless of his limited success in French politics, a source of inspiration for social and housing reformers all over Europe.[60]

Le Play's basic innovation in social research was to get away from statistical description and survey research and use a monographic treatment of family units. He analyzed household economics (mainly consumption) and more-or-less extended kinship ties in all parts of Europe for their effectiveness as socializing institutions and the backbone of social security.[61] Coming from a broken family himself and living through a series of revolutions and mounting urban poverty, he became increasingly obsessed by the idea of national decay and social instability as two sides of the same coin. His more famous remedies, the stem family and the paternalistic industrialist, may not have been very realistic. His perception, however, of the family as the basic social unit and an autonomous stabilizing force against the hazards of economic and political change (which through legislation as well as education should be handed down to the proletarian and in general become the cornerstone of social reconstruction[62]) proved important, because it seemed consistent with the evolving bourgeois family pattern and offered to the proletariat an immanent, or secular, alternative to revolution. It is true that Le Play's defense of the family was intermingled with references to eternal religious values, but his attitude was neither Christian nor clerical; religion for him serves the reestablishment of paternalistic authority and the enforcement of his aims, at which he arrived by more secular reasoning. His monographic family studies were aimed at singling out a basic social unit, designed to combine autonomous stability with achievement-oriented flexibility. Social reform, then, should restructure society so as to favor the spread of this model, and Le Play proved to be remarkably open and unprejudiced concerning the social partners, political means, and economic principles he chose in pursuit of this aim.[63]

The importance of Le Play does not rest with his model, the stem family, which his disciples had already devalued by showing it to be a transitory product of a destabilized patriarchal kinship organization; it derives instead from his approach to social reform, focusing on social microorganisms and building his program on the analysis of their inner workings and economic efficiency. Apart from religion, Le Play said, individual property and the family were the bases of society; the stem family provided for individual happiness and the growth of state power, and housing was central for both: "The home [*foyer domestique*] is the property par excellence and the permanent center of family feelings."[64] He goes on to define housing as the essential demarcation of the social units, without which the organism of the family would be open to all sorts of disrupting outside influences and could not work: "One of the fundamental agreements of all civilization is the *complete isolation* of each family's habitation."[65] Each house must provide for the separation of generations and sexes and assign to every reproductive activity its functional space. A garden, stables, and shops may provide for supplementary family income in addition to wages or other professional income. The amount and refinement of individual furniture, underwear, and clothing stimulate and strengthen the sense for self-respect, thrift, providence, and property. Last but not least, it is essential that the family house should be owned and kept in the family's possession from one generation to another. As a material representation of family identity, housing property reinforces the long- term stability of the basic social unity. In addition, after praising Mulhouse and similar paternalistic experiments to facilitate the accession of workers to housing property, Le Play pointed to even wider perspectives: "Amidst an improvident population, the passion of property has created a powerful enthusiasm for thrift. Moreover, the workers, who become proprietors, now understand the danger of political agitation, since they think only of ameliorating their status and of rising, as far as their abilities allow, into the ranks of the bourgeoisie."[66]

There remained, however, two or three economic problems: the middle class liked bourgeois behavior, but it needed workers. In fact, the chances were only slight that all workers could rise into the middle class by owning their houses, not only because they were "less disposed to conquer by work and thrift at other kinds of property"[67] (i.e., capital), but because even 5 percent philanthropy was only rarely attractive to the middle class at large in a time of much higher interest rates in other fields, and the first payment (in Mulhouse some 400 francs) was an insurmountable barrier for most workers. So Le Play turned to the Emperor to pave the way to housing property for the working class, suggesting that he would thus found his dynasty in social peace; but he ended somewhat helplessly with the all too pertinent remark, "The rigorous application of the economic principle of offer and demand disorganizes social relations in the field of rents as well as wages."[68]

By now, we may have gathered that economics was not our reformers' forte, since Le Play's cry for Imperialism or Paternalism was not really answered in his own days; possible exceptions were parts of the coal and steel industry, which had to build workers' housing anyway, being located in areas with no infrastructure and housing market, but at the same time forced to attract and hold a labor force. Given these conditions, some of Le Play's advice was heeded, especially his emphasis on isolating families and making housing property accessible.[69] But the usefulness of the family was not lost by a lack of economic realism, because its implications were much wider than the scope of Le Play's insufficient instruments. In a later tract he wrote, "Our most fatal error is the disorganization, by the disrespectful intrusion of the State, of the paternal authority in the family, the most natural and most fruitful of autonomies, which best preserves the social good, by repressing...corruption and orienting the youth to respect and obedience." The State and the bad instincts of the youth would lead society to a "savage state"; on the other hand, the reestablishment of fatherly authority would, "step by step, ...reestablish peace, together with respect and obedience, in private life as well as in local and central government."[70]

But this aim of peace should not be misunderstood as stagnation; it meant, rather, national or even racial unity "in the eternal competition for predominance" with patriarchal empires like Russia or China, with "their religious, fertile and docile races."[71] France's decadence, Le Play felt, was only the most extreme example of Western decay, indicated by sinking birthrates, a lost cultural leadership in Europe, and the lack of institutions to breed and select the best. These concerns were very much in his mind in advocating the stem family as a combination of a stable retreat with a production of some twenty children per marriage, the keeping of those unfit for reproduction in celibacy or servitude, and the identification of the fittest. He pointed to the example of the breeding and rearing of domestic animals: "They select with special care the individuals charged with the reproduction of each species. They, then, provide the young pupils [élèves] with the best means of development: a sane habitation, good nourishment, and the apprenticeship in the work needed by the master."[72]

It should be noted that the isolation of the family as an essential instrument of moral education and as a network of social security had played a prominent role with many social reformers before Le Play, as for example with Chadwick, but their approach was more naive in taking the functions of the family for granted.[73] Le Play brought these approaches together, constructed a typology of families and their role in social stability, and topped his synthesis with an articulate theory of the socializing functions of family roles and environments, of the necessity to restore family autonomy by political means, and of its sociobiological uses. Though here incorporated into the framework of the social doctrine of Catholicism, there were two principles, tinged with social Darwinism, that would later be elaborated by many bourgeois reformers and, most prominently, by the Fascists:[74] First was the selection and support of the fittest, which I shall leave aside here because the stem family never materialized and the idea was less influential on housing, besides the so-called Heimstaetten- Bewegung. More important was what I should like to call biological transcendency.

One of the essential problems of nineteenth-century class society was its frustration, in the midst of a society aimed at personal and general progress, of any prospects for the lower classes, short of fracturing the basic social structures defining the relations of the various classes to the means of production. Le Play's family theory provided an instrument to define a future, not relative to class experience, but relative to smaller social units, focusing on their reproductive and socializing capacity along the middle-class family structure. Lacking its base of capital and sophisticated education, the identity of these prospective family units among the working population had to be fostered by favorable circumstances[75] and the access to a gradual amelioration of consumption. Thus, the social ecology of housing, to be designed, subvented, and selectively distributed by social or public institutions to those who had already demonstrated their basic capacity for acculturation, began to rank high on the priority list of social control reform. Yet, housing proved to be a very expensive compensation.

A POSTSCRIPT ON THE
PERSPECTIVE OF REALIZATION

Though the housing reform debate by no means ended by the 1870s (in fact, the paper plans only increased in the ensuing decades), I shall stop here, because there now prevailed the new paradigm that authors like the ones I have quoted had constructed in the middle of the century. The problem was to translate aims into programs, make them concrete and feasible, and find a social agent to push them using pressure and power politics. This too is a fascinating part of the discussion, opening up the dimensions of the economics of housing and the politics of planning, but to elaborate on it goes beyond the scope of this essay.[76] Rather, I should like to conclude with a few remarks on five points that I think were basic for the development of the new paradigm into a full-fledged housing policy, realized in most parts of Europe around World War I: (1) the economic dilemma of the most expensive consumer commodity as an agent of reform; (2) the challenge posed by the spread of socialism at the end of the century; (3) the pressure of the lower-middle class for social space; (4) the implementation of working-class housing into city planning designs; and (5) the impact of war.

In his pamphlet on the housing question, the tract with the highest circulation among his writings at the time, Frederick Engels painted, in 1872, an ironic picture of the ups and downs of the housing reform debate, using Emil Sachs, a Viennese reformer, as a very apt example.[77] Engels showed how the aspiration of a housing reformer kept rising because of his humane concern, his social proposals, and his overall aims for a new and peaceful society; then, coming against the reality of economic feasibility, the author was quickly brought down to earth again, because his plans did not work under the conditions of bourgeois society, whose structure they were designed to preserve. This tendency is to be found almost everywhere, where housing reform was to transcend the isolation of philanthropy for support. Capitalist society did not work by insight but by profit. When the liberal and conservative housing reformers—who in the 1890s set up in France La Société des Habitations à Bon Marché—came down to a pragmatic program, they left all their former ideas about individual, detached houses surrounded by gardens and owned by workers, and satisfied themselves with the isolation of families in the small dwellings of big tenement blocks, thus perpetuating the urban density they had hitherto fought. This continued to hold true until intervening state subsidies offset some of the economic interests in the housing market. There were only a few solutions to the economic dilemma of realizing the new strategy, and these solutions were tied to very specific preconditions. One was paternalism, but this was practiced on a quantitatively interesting scale only in two fields: by state governments interested in recruiting an especially loyal blue- and more often white-collar labor force, and by the coal and steel industry, influenced not only by its old tradition but also by its inability to use urban infrastructures and housing markets. The second possibility was preferred by the liberals: building societies or similar cooperatives. Before state intervention, the cooperative movement on the continent was an utter failure in the field of housing, because the combination of the continental freehold system with extremely high land values and a more expensive and technically stagnating building production prevented, in contrast to England, even the petty bourgeois, the employee, and the foreman from owning housing property. A third was municipal socialism (i.e., experimental forms of state intervention by social or municipal corporations of self-government) to facilitate by a variety of regulatory, financial, and planning instruments cheaper, healthier, more nature- and family-oriented buildings for the lower half of the consumer population. All these approaches, though important in the evolution of models, could not be enlarged into a generalized solution, because the reframing of social space began with a substantial subvention to supply more dwelling space to those in the lower wage bracket, after experimenting and discussing through various decades, as those who had posed the question of working-class housing in the interest of the middle class came to realize that they stood in opposition to the basic economic control exercised by that same class.

Help in the transformation of ideas and plans into a workable policy came in the last two decades of the nineteenth century increasingly from two sides, the rise of new socialisms and the rise of a new lower-middle class, the officials, teachers, engineers, and office and commercial workers. The first must not be given too much importance, since most of the growing socialist movements in Europe had very little to say in the field of housing. Fixed on the workers' relation to the means of production, they lacked, despite the wealth of former utopias, a programmatic grasp of the problems of reproduction, and an alternative to bourgeois reform. But there were needs for space on the side of the working population even in the prerevolution era, which could become instrumental in bourgeois reform working as an incentive to individual adoption, apart from the lack of organized pressure. On the other hand, the growth of the socialist movements, especially when the high tide of strikes coincided with the decline of the building market, supported the idea that preventive social policy was needed, even at the risk of growing state power and a squeeze on private profits. But the preventive proposal to destroy the cultural preconditions of socialism by sociospatial strategies was not sufficient to establish a continuous and widespread policy, to which the interests of a similarly widespread class, that of the proprietors, would have had to yield; it provoked, however, scattered experiments and legislative projects, which were either barely enforced or enjoyed a very long sojourn in parliament and in the bureaucracy (in the case of Prussia, for more than two decades).

Perhaps more effective was the supplementary pressure from the lower-middle class, which was not able to escape from the growing urban density and immediate experiences of class mixture, as were the big bourgeois families. They began to form pressure groups like the German association for *Bodenreform*, putting planning politics and state subventions to reduce building costs high on the priority list of their respective political parties. They thereby enlarged the whole issue beyond the realm of economics into a genuinely political task: from dwellings to settlements and whole suburbs; from semidetached dreams to garden-city suburbs, implying a reconstructed mutliclass environment beyond the threat of big urban masses, to be brought about by state power; from the functionalist and symbolic division of spaces within an apartment to the zoning of the city. In this endeavor, a tiny but most influential fragment of socialism joined in (Fabians, Possibilists, Revisionists), representing the technocratic skill of socialist intellectuals and the acculturated needs of the white-collar working class.

The final transition from state intervention by regulation to one by subvention, to break the economic deadlock on the realization of sociospatial strategies, was in most European countries due largely to World War I. Then the building market broke down, and with it, at least for a short time, the legitimacy and efficiency of the proprietors' provisions for popular housing. The enemy was weak, and the mounting class tensions during the war persuaded the government that a more radical turn to preventive social policy was needed to preserve the essential structures. These tendencies were buttressed by very emotional petty bourgeois and white-collar pressures, which formed in most European countries some variety of a "Homes for Heroes" organization, agitating violently for a compensation for the bodily insults and the loss of status in a sheltered future. Wherever the working-class organizations came to exercise power in the early postwar days, they added their pressure to provide net dwelling space, the realization of preconceived political strategies of structural acculturation, which had, long ago, been thought up to destroy their threat.

NOTES

1. See, for instance, classics like Heinrich Herkner's *Die Arbeiterfrage,* 2 vols. (Berlin u. Leipzig, 1921), or more recent treatments, like Maurice Bruce's *The Coming of the Welfare State* (London, 1961), Albin Gladen's *Geschichte der Sozialpolitik in Deutschland* (Wiesbaden, 1974), or Patrick de Laubier's *L'âge de la politique sociale* (Paris, 1978).

2. Enid Gauldie, *Cruel Habitations* (London, 1974); Roger-Henri Guerrand, *Les Origines du logement social en France* (Paris, 1967).

3. See, for instance, Stanley D. Chapman (ed.), *Working-Class Housing* (Newton Abbot, 1971); Anthony Sutcliffe (ed.), *Multi-Storey Living* (London, 1974); Michael A. Simpson and Terence H. Lloyd (eds.), *Middle-Class Housing in Britain* (Newton Abbot, 1977).

4. To choose German dissertations for example this time, see Adolf F. Heinrich, *Die Wohnungsnot und die Wohnungsfuersorge privater Arbeitgeber in Deutschland im 19. Jahrhundert* (Marburg, 1979); Kristina Hartmann, *Deutsche Gartenstadtbewegung* (Muenchen, 1977); Dorothea Berger-Thimme, *Wohnungsfrage und Sozialstaat* (Frankfurt, 1976).

5. Again just examples: William Ashworth, *The Genesis of Modern British Town Planning* (London, 1954); W. Vere Hole, *The Housing of the Working Classes in Britain, 1850-1914* (London University Ph.D., 1965); J.N. Tarn, *Working-Class Housing in 19th-Century Britain* (London, 1971).

6. This term, to be explained below at the example of Le Play, is meant in the sense of the philosophic concept of transcendency—as a pacified (familistic and biological), alternative future for the working classes—as against the collectivist transcendency of socialist perspectives.

7. The term "sociospatial control" is not used here as an analytical-historical concept of functionalist totality within a three-stage model (of traditional society, chaos and class struggle, and reestablished social control by the *embourgeoisement* of the masses), as it was first styled by the conservative sociologist E.A. Ross in *Social Control: A Survey of the Foundations of Order* (New York, 1929), and taken up by P.A. Landis, T. Parsons, and others. See the critic of Gareth Stedman Jones, "Class expression versus social control," in *History Workshop* 4 (1977), 162 ss. It seems, however, very suitable as a descriptive concept to organize the middle-class *intentions* behind social policy producing various programs that were difficult to translate into practice and, though the effectiveness of some of these cannot be denied, proved quite often partially counterproductive. Housing seems to be a very good case for these ambiguities. It is all the more astonishing to see an interesting collection of essays (A.P. Donajgrodzki, ed., *Social Control in 19th-Century Britain,* London, 1977) concentrate on police, leisure, education, religion, and social work and neglect the sociospatial and physical dimension almost completely.

8. It should be stressed that Chadwick, Hobrecht, and Le Play were not chosen as heroes of a *Geistesgeschichte* of housing reform, but as characteristic and especially

well articulated examples of the many contributors to the debate. I am in this chapter not interested in the intellectual genesis of certain reform programs, but in the contexts, perceptions, and interests that structured the contributions to the debate on sociospatial control.

9. See Gareth Stedman Jones, *Outcast London* (Oxford, 1971).

10. See numbers of the French series "recherches" like Lion Murard and Patrick Zylberman, *Le petit travailleur infatigable* (No. 25, 1976); Isaac Joseph and Philippe Fritsch, *Discipline à domicile* (28, 1977). *L'Haleine des faubourgs* (29, 1977), or the contributions to Lutz Niethammer (ed.), *Wohnen im Wandel* (Wuppertal, 1979).

11. The following model cannot be footnoted to represent all the researches and experiences that informed it, because references would either drown the text or be unessential. For bourgeois culture I should like, however, to point to the interpretations of the Elias School. See Norbert Elias *Ueber den Prozess der Zivilisation*, 2 vols. (Frankfurt, 1976; New York, 1978, *The Civilizing Process*); Peter Gleichmann et al. (ed.) *Materialien zu Norbert Elias Zivilisationstheorie* (Frankfurt, 1979).

12. See Fernand Braudel, *Capitalişm and Material Life, 1400-1800* (London 1973); Hans Medick, "The proto-industrial family economy," in *Social History* 1 (1976), 291 ss.

13. For the latter aspect, doll houses are a good source. See, for example, Flora Gill Jacobs *A History of Doll Houses* (New York, 1953); Gottfried Korff, "Puppenstuben als Spiegel burgerlicher Wohnkultur," in Niethammer (ed.), *Wohnen*, 28 ss.

14. The change in approach to the bodily existence has become an important object of discussion mainly in France. See, besides notes 10 and 11, Francoise Loux, *Le corps dans la société traditionelle* (Paris, 1979); Jean Paul Aron and Roger Kempf, Le pénis et la démoralisation de l'Occident (Paris, 1978).

15. A good example for this are the houses of the most prominent nineteenth-century housing reformers, who almost invariably owned, built, bought, or rebuilt castles, chateaux, or major country or suburban cottages with a wealth of functional and symbolic inner differentiation and set into the environment of a tamed and unused sort of nature at good distance from urban mobs.

16. Besides the familiar literature on suburbanization and garden cities—at its best, H.J. Dyos and D.A. Reeder, "Slums and Suburbs," in H.J. Dyos and Michael Wolff (eds.), *The Victorian City,* 2 vols. (London, 1973), vol. 1, 359 ss.—see the more general discussion of the underlying motives in Pierre Sansot et al., *L'éspace et son double* (Paris, 1978).

17. The most intense set of quotations from such middle-class perceptions of the proletariat (in the early twentieth century in Germany) is to be found in Klaus Theweleit, *Maennerphantasien,* 2 vols. (Frankfurt, 1977). Earlier evidence is abundant in works like Louis Chevalier's *Labouring Classes and Dangerous Classes* (London, 1973; French: Paris, 1958); Louis Cazamian, *The social Novel in England 1830-1850* (London 1973); Dyos/Wolff (eds.), vol. 2, Parts V and VI; Anthony S. Wohl, *The Eternal Slum* (London 1977); Stedman Jones, op. cit.; Murard/Zylberman, op. cit.

18. On the amount and meaning of mobility and poverty, see *Chevalier*, op. cit.; Raphael Samuel, "Comers and Goers," in: Dyos/Wolff (eds.), vol. 1, 123 ss.; Otto Ruehle *Illustrierte Kulturund Sittengeschichte des Proletariats*, vol. 1 (Frankfurt, 1971), vol. 2 (Giessen, 1977); Dieter Langewiesche, "Wanderungsbewegungen in der Hochindustrialisierungsphase," in *Vierteljahresschrift füer Sozial- und Wirtschaftgeschichte* 64 (1977), 1 ss.; Franz Brueggemeier "Soziale Vagabundage oder revolutionaerer Heros," in Lutz Niethammer (ed.), *Lebenserfahrung und Kollektives Gedaechtnis* (Frankfurt, 1980), 193 s. The following sketch is inspired by these and similar contributions and backed by a rereading of social surveys by housing reformers

and other contemporary observers of the making of the industrial and metropolitan urban societies.

19. In using this concept (cf. Oscar Lews, "The Culture of Poverty," *Scientific American* 214, 1966, 19 ss.), I accept what has been concluded from a critical review of Mayhew (Gertrude Himmelfarb, "The Culture of Poverty," in Dyos/Wolff, vol. 2, 707 ss. quotation 731: "Then, as now, what passed as 'the culture of poverty' was the culture of a small subgroup of the poor. Yet that culture, or more precisely the image of that culture, was of momentous consequence in shaping the lives not only of the poor but of society as a whole." On the other side, this remarkably stable subgroup shared many features with other strata of the urban proletariat, especially on the Continent, so that the generalizations of bourgeois anxieties should not be debated away as if they were mere projections.

20. For a discussion of this concept, see J.K. Modell and T.K. Hareven, "Urbanization and the Malleable Household," *Journal of Marriage and Family* 35 (1973), 467 ss.; L. Niethammer and F. Brueggemeier, "Wie wohnten Arbeiter im Kaiserreich?" and *Archiv fuer Sozialgeschichte* 16 (1976), 61 ss.; Josef Ehmer, "Wohnen ohne eigene Wohnung," in L. Niethammer (ed.), *Wohnen*, 132 ss.

21. This aspect of collective socialization is more developed in Brueggemeier, *Soziale Vagabundage,* op cit.

22. Boissy d'Anglas in the session of the Assemblée of messidor 5, year III (Moniteur, messidor 12). This and the following two quotations are taken from Roger-Henri Guerrand, *Origines*, 18 s. For an even more useful pathfinder to similar material, see Roger-Henri Guerrand, *Le logement populaire en France: Sources documentaires et bibliographie (1800-1960)* (Paris, 1979).

23. *Journal des Débats,* 18.4.1832.

24. Dr. Taxil, *Topographie physique et médicale de Brest et sa banlieu* (Paris, 1834).

25. See Cazamian, op. cit.

26. Klaus-Michael Bogdal *Schaurige Bilder. Der Arbeiter im Blick des Burgers* (Frankfurt 1978); and note 17.

27. See note 24.

28. See Chevalier, op. cit., for Paris and for Berlin, e.g., Hartmut Frank "Wilhelm Stier: Schizzo per una Armenstadt, Berlino, marzo 1848," in P. Morachiello and G. Teyssot (eds.), *Le machine imperfette* (Roma, 1980), 452 ss. For the persistence of this social segment and its remobilization in the 1980s, see Stedman Jones, op. cit.

29. Examples for this equation are abundant. Here, one somewhat more sophisticated, from a famous conservative social reformer: "Like some of the Spanish Colonists in Southern America, who were cut off and relinquished in the virgin forests, sank completely back to the cultural level of the Indians, so our society forces the lower strata of the industrial proletariat of the big cities to sink, with the utmost necessity, back to a level of barbarity of bestiality, of brutality and hooliganism, that our ancestors had left behind since ages. I want to assure that the biggest danger menacing our civilization comes from here. The doctrines of the Social Democracy and of Anarchism only become criticial, when inseminated into a soil that is so dehumanized and horrible" Gustav Schmoller: Ein Mahnruf in der Wohnungsfrage, in: *Schmollers Jahrbuch 1887*, p. 429 s.

30. See note 28; Jeanne Gaillard, *Paris, la ville 1852-1870,* (Paris, 1977); Guerrand, *Origines,* 83 ss., 183 ss.; Horant Fassbinder, *Berliner Arbeiterviertel,* 1800-1918 (West Berlin, 1975).

31. Theodor Fritsch, *Die Stadt der Zukunft* (Leipzig, 1896), 7. It should be noted that Fritsch was also the author of a highly circulated antisemitic handbook, reprinted many times by the Nazis. But it would be premature to relegate him to the lunatic

fringe. Even though few of the pioneers of sociospatial policies may have shared his antisemitism, most of them joint in his basic thought about imperatives of urban structures and many experimented with political connotations of space and human biology.

32. See note 18.

33. B. Friedmann, *Die Wohnungsnot in Wien* (Wien, 1857).

34. For general biographical information, see the Introduction in Benjamin Ward Richardson, *The Health of Nations, A Review of the Works of Edwin Chadwick*, 2 vols. (London, 1887); Maurice Marston, *Sir Edwin Chadwick* (London, 1925); S.E. Finer, *The Life and Times of Sir Edwin Chadwick* (London, 1952).

35. Anthony Brundage, *The Making of the New Poor Law 1832-39*, (London, 1978).

36. See, for example, M.C. Buer, *Health, Wealth and Population in the early days of the industrial revolution* (London, 1926; reprinted 1968).

37. See R.A. Lewis, *Edwin Chadwick and the Public Health Movement*, (London, 1952), 29 ss.; M.J. Cullen *The Statistical Movement* in *Early Victorian Britain* (New York 1975), 53 ss.; Margaret Pelling,*Cholera, Fever and English Medicine 1825-1865*, Oxford 1978.

38. M.W. Flinn (ed.), *Report on the Sanitary Condition of the Labouring Population of Gt. Britain by Edwin Chadwick*, (Edinburgh, 1965).

39. Ibid., 252.

40. Ibid., 266 s.

41. Ibid., 268.

42. It seems quite consequent that Chadwick's sanitary writings were edited by his disciple under a title alluding to Adam Smith (see note 34).

43. Richardson (ed.), *Health of Nations*, vol. 2, 263 ss., 288-291, 293 s., 297 ss., 302-303.

44. Ibid., 304-306.

45. Ibid., 299-302.

46. James Hobrecht, *Ueber offentliche Gesundheitspflege und Bildung eines Central-Amts füer oeffentliche Gesundheitspflege im Staate* (Stettin, 1868), 12.

47. Water was very different in this respect, because its circulation called for public investment of the greatest order. See, for the most famous example of that time, *Mémoires du Baron Haussmann, Grand travaux de Paris*, 2 vols. (new edition, Paris, 1979), vol. 2, 7-151. References for continental discussions about sewage and water closets are in Peter Reinhardt Gleichmann, "Die Verhaeuslichung koerperlicher Verrichtungen," in Gleichmann et al. (ed.), *Materialien*, 254 ss.; and "Staedte reinigen and geruchlos machen," in Hermann Sturm (ed.), *Aesthetik und Umwelt*, (Tuebingen, 1979), 99 ss.

48. Best described in a curious utopia by B.W. Richardson, *Hygea, a City of Health* (Brighton, 1875), dedicated to Chadwick.

49. See E. Heinrich, "Der Hobrechtplan," in *Jahrbuch fuer Brandenburgische Landesgeschichte* (Berlin, 1962); D. Radicke, "Der Berliner Bebauungsplan von 1862 und die Entwicklung des Wedding," in *Festschrift E. Heinrich* (Berlin, 1974).

50. Cf. David H. Pinkney, *Napoleon III and the Rebuilding of Paris* (London, 1958); Louis Girard, *La politique des traveaux publics du Second Empire* (Paris, 1952); Anthony Sutcliffe, *The Autumn of Central Paris* (London, 1970).

51. This point of visibility may be also found in Chadwick, see A.P. Donajgrodzki, "'Social Police' and the Bureaucratic Elite: A Vision of Order in the Age of Reform," in Donajgrodzki (ed.), 51 ss.

52. All quotations from Hobrecht, 12-16.

53. Among the critics from the ranks of mainstream bourgeois housing reform are most noteworthy: Ernst Bruch, *Berlins bauliche Zukunft,* (Berlin, 1870); Rudolf Eberstadt, *Handbuch des Wohnungswesens und der Wohnungsfrage* (2nd ed. Jena, 1910), 240 ss.; Werner Hegemann, *Das steinerne Berlin* (1st ed., 1930; 2nd ed., Berlin, 1963), 207 ss.

54. The one great exception in mid-nineteenth-century urbanism who got beyond utopia into the practice of a progressive technocrat, based on social and spatial science of urbanism, seems to be Il defonso Cerdà, in *La theorie générale de l'urbanisation* (abridged translation edited by Antonio Lopez de Aberasturi; Paris, 1979; 1st ed., 2 vols., Madrid, 1867); Arturo Soria Y Puig, *Hacia una teoria general de la urbanization* (Madrid, 1979); Juan Rodriguez-Lores, "Il defonso Cerdà: Die Wissenschaft des Staedtebaues und der Bebauungsplan fuer Barcelona 1859," in *Stadtbau um die Jahrhundertwende* (Koeln, 1980). See also his interesting comparison of the expansion schemes of Hobrecht and Cerdà, "Die Grundfrage der Grundrente," *Stadtbauwelt* 65 (28.3.1980), 443 ss. But admitting the superiority of Cerdà's analyses of the social heterogeneity of the city and of the functions of its life, should not preclude some basic correspondences with other bourgeois reformers: his planning is inspired by the wish to restructure and improve the public and private life of the working population without basic changes in the sphere of production, and he conceives of the private room as the *sanctum sanctorum.* According to Cerdà, the three principles of future urbanization should be public health, urban traffic and a just policy of real estate. But his aproach to city expansion by the imposition of a leveling, rectangular street plan could not prevent land speculation with the result of high-density blocks and residential differentiation.

55. See Friedrich Mielke, "Studie ueber den Berliner Wohnungsbau zwischen den Kriegen 1870/71 und 1914-1918," in *Jahrbuch füer die Geschichte Mittel- und Ostdeutschlands* 20 (1971), 202 ss., for an interpretation of the *Mietskaserne* as a mere blow-up of preindustrial building types. More balanced views appear in Manfred Hecker, "Die Berliner Mietskaserne," in Ludwig Grote (ed.), *Die deutsche Stadt im 19, Jahrhundert* (Muenchen, 1974), 273 ss.; D.R. Frank and D. Rentschler (eds.), *Berlin und seine Bauten,* part IV, vol. A (Berlin, 1970).

56. Taking Emil Sax, *Die Wohnungszustaende der arbeitenden Classen und ihre Reform* (Wien, 1869), as an example, this was criticized in a classic way in Friedrich Engels, "Die Wohnungsfrage," in *Marx-Engels-Werke,* vol. 5 (East Berlin, 1959), 210-287.

57. This communication network has not yet been sufficiently researched, though some planning and architectural histories reproduce its international approach. See Leonardo Benevolo *History of Modern Architecture,* 2 vols. (London, 1971); Michael Ragon, *Histoire mondiale de l'architecture et de l'urbanisme moderne,* 3 vols. (Paris, 1971). Within his long-term project on the problem of working-class housing in nineteenth-century Europe, the author hopes to contribute to the reconstruction of this international community of sociospatial reform.

58. See note 6.

59. For general biographical information, see the good summary by Jesse R. Pitts, in *International Encyclopaedia of the Social Sciences,* vol. 9, 1968, 84-91; Michael Z. Burke, *Le Play: Engineer and Social Scientist* (London, 1970). The only modern text collection seems to be Louis Baudin (ed.), *Frédéric Le Play, 1806-1892* (Paris, 1947). Interesting is the introductory biographical sketch by one of his disciples Charles de Ribbe, *Le Play, d'après sa correspondance* (Paris, 1884).

60. For France see Guerrand, *origines,* 257 ss.; for Germany e.g. M.W. Roscher, *Gescnichte der Nationaloekonomie in Deutschland,* (Muenchen, 1875); A.E.F. Schaeffle, *Das gesellschaftliche System der menschlichen Wirtschaft* (Tuebingen, 1867).

His most important impact on modern urbanism seems to have been channelled through Geddes, whose system of "place, work, folk" echoed Le Play's formula "lien, famille, travail." V.V. Branford and Patrick Geddes, *The Coming Polity* (London, 1919); Philip Mairet, *Pioneer of Sociology—the life and letters of Patrick Geddes,* (London, 1957); Philip Boardman, *The Worlds of Patrick Geddes,* (London, 1978).

61. Frédéric Le Play, *Les Ouvriers européens,* 6 vols. (Tours, 1877-1879). He walked all over Europe and seems to have been a communications genius in establishing contact with farmers, artisans, and the like in all languages, but with the urban proletariat. When entering the East End of London in the 1850s, he enjoyed the company of four policemen in meeting this "dangerous population of blacklegs, thieves, murders, and corresponding women" in their "lodging houses with a dozen beds with twenty people of all ages and both sexes," who were "really curious to look at" (Burke, 12).

62. Frédéric Le Play, *L'Organisation de la Famille* (Paris, 1871).

63. He progressed from a temporary socialist in 1848 to a consolidated conservative, manager, and adviser to Napoleon III, but he also advocated anticapitalist measures and became one of the involuntary fathers of the Socialist International by persuading French authorities to send a workers' delegation to the London World Exhibition in 1862 (Burke, 61).

64. Frédéric Le Play, *La Réforme Sociale,* 2 vols. (Paris, 1864), vol. 1, 170.

65. Ibid., 175.

66. Ibid., 172 s.

67. Ibid., 173.

68. Ibid., 174.

69. See Murard/Zylberman, op. cit.

70. Le Play, *Organisation,* XVI-XX.

71. Ibid., IX.

72. Ibid., 3. This is one of the first sentences of §1 on "progress and stability of races."

73 See his chapter, "The Want of Separate Apartments, and Overcrowding of Private Dwellings" in *The Sanitary Report,* Flinn (ed.), 190 ss.

74. For some of the assumptions of housing reform that could be taken up by Fascism, see Klaus Bergmann, *Agrarromantik und Gross stadtfeindschaft* (Meisenheim am Glan, 1970); Lion Murard and Patrick Zylberman "La Cité eugénique," in *L'Haleine des faubourgs,* 423 ss.

75. This familistic approach did, in the long run, however, not work under the conditions of urban agglomerations, but only in relatively isolated industrial villages. See the perfect little paper of Lucian Brams, "Structures sociales et famille ouvrière," in *Transactions of the 3rd World Congress of Sociology* (London, 1956), vol. 4, 146 ss.

76. Regarding the summarizing character of this postscript, I refrain from giving further references, which would be bound to overflow the text. Looking to the German example, I have tried to give a more detailed account of the coming of state intervention in the housing market in the article, "Ein langer Marsch durch die Institutionen: Zur Vorgeschichte des preussischen Wohnungsgesetzes von 1918," in Niethammer, *Wohnen im Wandel,* 363 ss.

77. See note 56.

6

COMMENTS

Christine M. Rosen:

The preceding two chapters concern a very important issue: the use of housing as a mechanism of social control. They concentrate on very different aspects of this problem. On one hand, Professor Niethammer has addressed the issue from the viewpoint of explaining the goals and motivations of nineteenth-century European housing reformers, and has attempted to ground his analysis within a larger analysis of the nature of industrial capitalism and what he has called the cultural dialectic of urban society.

On the other hand, Professor Jackson has more or less picked up where Professor Niethammer left off and concentrated on analyzing the housing policies themselves. His concern is with the twentieth century, of course, rather than the nineteenth, and while he has had made certain conclusions about the goals and motivations of the policy makers, he has focused his energy on explaining the technicalities of housing programs like the HOLC and the FHA, in order to make clear why the programs had the impact on suburbanization and inner-city decline that they had. One of his major conclusions is that the programs severely damaged ethnic and racial minorities and the poor in general. Unlike Professor Niethammer, however, he has not attempted to fit his analysis of the programs and their impacts into a larger analysis of industrial society.

These are *very* different perspectives, and they may make the chapters seem rather incomparable, at first glance at least.

Despite their different focuses, however, they share two important common bonds that I think are worth discussing right at the outset. The first thing they share is, of course, the idea that housing is an important mechanism of social control, something that can be used by elites as a countervailing force against social disorder and class conflict. This is an extremely important point and gets to the root of some of the problems with traditional histories of housing reform that view reform purely in liberal, humanistic terms.

The two chapters also share an overall way of considering the issue of housing and social control. Despite their differences, both have taken a "from the top down" approach to the problem—treating the quantity, quality, and location of housing as instrumentalities that are manipulated by public policy makers for various social purposes. This is a perfectly valid and, as I just said, very important and necessary way of examining housing reform and policy. It is not, however, the only way of analyzing housing as a mechanism of social control and order.

What I would like to do in my comment is discuss each of the chapters on its own terms, and then briefly describe an alternative, more organic, "from the bottom up" approach to the housing and social control issue that draws on my own research into workers and working-class housing in Chicago in the 1870s. Afterwards, I would like briefly to reinterpret the chapters in light of that approach.

As the title of his chapter indicates, the heart of Professor Niethammer's study is his analysis of the paradigm of social control that developed in the course of the housing reform and general social reform debates that took place in Europe in the mid- and late nineteenth century. Professor Niethammer argues that this paradigm had three basic components, which he has exemplified with analyses of the writings of Edwin Chadwick, James Hobrecht, and Frederick Le Play.

What is especially exciting about his essay, however, is that it is much more than a simple description of the components of a strategy or paradigm of social control. It is also a psychological interpretation of the reformers' ulterior motives and, more important, a Marxist attempt to place the housing debate within the larger context of the social and cultural changes that were occurring in European society at the time. In this regard, Professor Niethammer makes several interpretive points that I think make his study a very interesting and provocative contribution to the historiography of housing reform.

One such contribution is his psychological analysis of the reformers' motivations. He argues that people like Chadwick, Hobrecht, and Le Play were not responding to the misery of the people living in slums so much as they were responding to the inner turmoil that they themselves felt regarding the cultural changes produced by industrialization and the growth of cities, an inner turmoil that they projected onto the poor because the feelings of fear, revulsion, and chaos the slums inspired in them were so close to the feelings of fear, revulsion, and chaos that modernization inspired in them. It is, of course, difficult to document psychological arguments like this, but Professor Niethammer emphasizes the sensual perceptions of a whole culture at risk that emerge in the reformers' writings, to support his argument that they had culture and not the physical welfare of the masses in mind.

Another major point is Professor Niethammer's characterization of housing reform—which follows from this interpretation of the reformers' motivations. It is that housing reform was an instrument of cultural reproduction—in other words, an instrument of psycho-socio-cultural stability, rather than an instrument of social and environmental improvement, per se, or even an instrument of socioeconomic reform that one might expect the capitalist class to have initiated in order to enhance and stabilize industrial production. In his view, the reformers were guardians of culture who occupied a rather precarious middle ground between the completely alienated masses and the comfortable and complacent capitalist class. In other words, they were hardly the partners of the workers in their efforts to improve the workers' lot; at the same time, they were not the partners of the capitalists either, since they were trying to interfere with the status quo. This distinction has non-Marxist parallels in American history in the controversy over the status anxieties and social origins of Progressive reformers. It also helps flesh out Professor Niethammer's argument concerning the limited nature of housing reform. He contends that housing reform failed to eliminate slums because it was an effort by middle-class people to deal with their own inner conflicts over the modernization of society, by devising ways to maintain middle-class cultural norms and promote the norms' acceptance by the working class—rather than a true effort to help workers by treating the slum problem at its source through a transformation of the economic structure of society.

This is, I think, a new interpretation of reform that raises all sorts of questions worth discussing, several of which I will raise here.

One question is whether Professor Niethammer's psychological interpretation of the housing reformers' motivations provides a complete explanation for the existence and content of the housing reform movement: Was housing reform really *nothing* more than a psychological defense by neurotic middle-class people who could not reconcile themselves to the cultural and social chaos and confusion caused by the industrialization process? Or did it have other causes?

A related question is wheter it is accurate, or even all that desirable, to argue that the reformers care nothing about the sufferings of the poor, but were really only interested in using housing reform to reproduce middle-class values in the working class. Admitting that at least some of these reformers cared somewhat about the suffering of workers would *add*, I think, to our understanding of the irony of the reformers' halfway position on reform and the paradox of and their failure to achieve the elimination of the slums, the complete elimination of the working-class threat, or the end of the continuing cultural chaos caused by the ongoing modernization process.

Another question is whether it is fair to criticize reformers for not viewing the housing problem in the same terms as the people living in the slums would have. Running through this essay is the idea that because reformers focused in their writings on the social costs of slum housing, they were responding to the problem in a peculiarly middle-class way. This is an argument that to me, at least, reads as a criticism of the reformers rather than as a bald statement of fact, the implicit assumption being that they ought to have viewed the slums as the poor did. The fact is, however, that because slums are inhabited by the poor, the misery they generate is largely internalized by the poor. Other people may feel pity or even try to empathize, but short of joining them, they can only directly suffer from the problem by experiencing the negative externalities and social costs the slums cause. For what it is worth, I think this means that in a *class-divided* society, the perception of these negative externalities and social costs may well be the most authentic way in which middle-class people can identify themselves with the housing problems of the poor.

As I said, Professor Jackson picks up where Professor Niethammer leaves off and examines actual housing programs and their impact on society. His chapter is an ambitious study that covers not only the fairly well-known story of the origins, problems, and negative impacts of federal urban renewal and public housing policies, but also the less well-known beginnings, flaws, and negative impacts of various federal mortgage assistance programs.

He traces the development of these financial programs from the establishment of the Home Owners' Loan Corporation in 1933 through the establishment and development of the FHA and Veterans Administration housing programs. His maps and graphs of the HOLC's and the FHA's mortgage activities in various counties vividly document the ways these programs discriminated against reinvestment and new investment in black and ethnic neighborhoods, as well as the way they favored the construction of single-family homes in the suburbs at the expense of multifamily housing in the inner city.

Professor Jackson's analysis of the antiblack, anti-ethnic, anti-urban nature of the programs is very convincing, I think. The question I would like to raise is whether it is entirely valid to argue that the policy makers and bureaucrats who created the programs did so with the intention of using them as instruments of social control. Clearly, the programs did have significant social and spatial impacts and did discriminate against certain groups and areas. The question, though, is whether the discriminatory features of the programs were deliberately used as instruments of social control, or primarily intended to minimize investment risk. The discriminatory directives and requirements did, after all, reflect prevailing appraisal standards, and the programs were organized to pay for themselves over the long term.

The question boils down to whether the agencies' insistence on investment risk minimization was just a cover-up for the racist, anti-urban goal of creating lily-white, ethnically homogeneous suburbs, or whether it was a less insidious attempt simply to minimize risk on the basis of appraisal standards that were fundamentally flawed by the racism and the pro-suburb, pro-single-family homeism that permeated American society at the time. As Professor Jackson points out, the programs were established to stimulate homeownership and not to revive decaying cities.

The same question can be raised about urban renewal. Was it the *goal* of the policy makers who created the programs that they would use primarily to facilitate the construction of office buildings and highrise apartment buildings, or were the programs, because of technical flaws, simply taken over and used by private developers for their own purposes? This question is, I think, trickier to answer, since a great many conflicting motivations seem to have gone at least into the *writing* of the law, if not into the *idea* behind the program.

And this raises another, even bigger question. That is, what kind of social control were these programs actually promoting, anyway? I spent some time trying to fit them into Professor Neithammer's paradigm and decided that the *idea* of mortgage insurance fit in with his familistic Le Playan model and the *idea* of urban renewal fit in with his sanitary-functionist Chadwickian model. But the all-important discriminatory aspects of the mortgage programs and the all-important commercial bias of urban renewal did not fit into the paradigm at all. Neither could have been meant to help or to stabilize and absorb into the middle class the minority and working-class inhabitants of the slums, either politically, economically, culturally, or socially. Instead, the discriminatory features were clearly meant to do the opposite and insulate homeowners, CBDs, and the investment portfolio of the federal government from the blight of inner-city slums.

This leads me to suggest that it may have been the policy makers' *failure* to incorporate into their programs an effective mechanism of social control relating to the lower classes that paved the way for the inner-city riots of the 1960s. In other words, the policy makers may have *overlooked* the social control potential of housing policy in formulating the programs— at great cost to social stability. In this regard, it is interesting to note that after the riots, the government created new mortgage programs in the FHA, the Federal National Mortgage Association and the Federal Home Loan Bank Board, in the hope of funneling more mortgage credit into inner cities; the reason, of course, was to prevent future riots.

As I said, both Professor Niethammer's and Professor Jackson's studies are "top down" considerations of the housing and social control issues, focusing on the way elites tried to use housing as instruments of control. Again, this is an important way of looking at the social control issue, but it is not the only way of looking at it. It makes an assumption that bears examining; that housing conditions do, in fact, foster stability in ways that can be manipulated by power elites.

Now, American historians have not been as articulate about the manner in which they think housing affects social and cultural relationships as were reformers in nineteenth-century Europe. The prevailing view, however, seems to be that unhealthy and congested slum conditions actively encouraged social tension and conflict, while clean, multifamily housing was at least neutral, and homeownership positively inspired stability—by giving even poor and unskilled workers an avenue of social mobility that allowed them to identify themselves with the middle class. The question, however, is whether housing conditions really promoted and discouraged social harmony in this way. To answer these questions, I think it is necessary to examine the way workers, themselves, reacted to and viewed homeownership and housing conditions in general. My research into workers and working-class housing during the reconstruction of Chicago after the Great Fire of 1871 enables me briefly to suggest some answers to the question, at least insofar as homeownership is concerned.

I have been examining the environmental improvements Chicagoans attempted to institute in the aftermath of the Great Fire. One such improvement was an attempt to eliminate all wooden construction from the city. It resulted in an ordinance to outlaw all frame construction that produced a confrontation between the city's workers and the middle class. That confrontation revealed a great deal about the workers' attitudes toward their homes, as well as a great deal about class tensions in the city in general.

Because the Great Fire had started in a wooden barn and had been fed by the wooden buildings that filled the city, most middle- and upper-class people saw the ordinance as an essential reform that was needed to prevent future conflagrations and hold down insurance rates. Many liberal reformers also saw it as a kind of social reform that would curtail the power of land speculators and slumlords by assuring the construction of working-class housing that would be not only more substantial but also cheaper to heat than the shanty housing the fire had destroyed.

To the city's immigrant workers, however, the ordinance was nothing less than an attempt to prohibit them from continuing to build and own their own homes. They had always owned a large part of the city's housing stock, partly because land outside the city center was cheap, partly because big landowners leased plots of land and allowed them to build their own structures, saving them land costs, and partly because the city had never before done anything to stop them from constructing wooden shanties and cottages, the only kind of housing they could afford to build.

In the space allocated to me, I cannot give the full story of how working-class homeowners and would-be homeowners attempted to stop passage of the ordinance. The actions they took, however, show that they cared deeply about homeownership and that they saw the attempt to outlaw wood construction in class terms as an attack by the upper classes on their independence and material well-being. Among other things, they wrote petitions against the ordinance and in protest staged a massive march on city hall that ultimately disintegrated into a near riot. The preamble to a petition against the ordinance put their feelings this way:

> There are evil counselors who persistently urge upon our public authorities the assumed necessity of crushing out the poorer lot owners by preventing them from rebuilding in such a manner as their limited means enable them to do.

One resolution added,

> Instead of enjoying the blessings of independent homes, our laboring people would be crowded into those terrible tenement houses which are the curse of eastern cities. The effect of the great fire would then have been to make the rich richer and the poor poorer.

Speakers also emphasized the idea that the ordinance would, in their words, "crush out" the workers' lives. Most of the participants in the march on city hall were German, Swedish, Norwegian, native homeowners and would-be homeowners from the north side of Chicago. Many carried banners and, like the verbal rhetoric in the petitions, the banners' visual rhetoric revealed the class tensions that helped activate the protest. The slogans included phrases like "Leave a Home for the Laborer," "Don't Vote Anymore for the Poor Man's Oppressor," "The Voice of the People," "No Barracks," and "No Tenements." One banner even showed a gallows and a set of carpenter tools arranged to resemble a skull and crossbones, with the slogan, "This is the lot of those who vote for the fire limits."

Needless to say, middle-class people were horrified and outraged by the demonstration. What is significant, however, is that it gave the workers what they wanted—a fire ordinance allowing them to build wooden homes almost wherever they wanted—and that as a result, the overt conflict and tension between the classes quickly disappeared.

The existence and peaceful outcome of this episode clearly demonstrates that homeownership helped promote social stability. What is interesting, though, is that it shows that homeownership did not necessarily do this by permitting workers to identify with the middle class or by improving living conditions. These people plainly viewed themselves as workers. Furthermore, they built the same kind of small one- and two-room wooden cottages (or shanties, as they were often called) as speculators did, often on peripherally located and unsewered land plagued by miasmatic smells and poor drainage, usually renting out part of the limited amount of space in the houses to other workers. What the episode suggests about the relationship between housing and social control is that homeownership (and even the promise of homeownership) fostered stability by giving workers a sense of control over their living conditions that reduced tension by making the harsh conditions of working-class life more bearable and by giving the people, *as workers,* a sense of dignity and independence. The other side of this relationship, of course, is that threats to deprive workers of their ability to obtain their own homes could easily destroy the modus vivendi of the classes, bringing the underlying tensions between them to the fore.

This "from the bottom up" perspective on the social control issue helps expand on some of the issues raised by Professor Niethammer in his chapter. Clearly, this kind of social control is very different from the kind of social control through cultural reproduction envisioned by his reformers. In Europe, of course, working-class homeownership was not as prevalent as it was in America, where, in most cities besides New York City, workers owned a large share of the working-class housing stock. This suggests that Professor Niethammer's picture of completely alienated workers, deprived, in his words, of their "material civilization," probably did not hold true in the United States. This probably helps explain why American workers were not as restive as those in Europe. It may also be, however, that European workers were not quite as totally culturally deprived and oppressed as Professor Niethammer suggests. I know several historians and sociologists who have begun examining what they call the material culture of the working class and who have found that working-class families personalize their living environments in all sorts of ways that suggest a very important autonomous cultural dimension of working-class life, regardless of whether they own their own homes. This may have been another, probably weaker, source of social stability in both the United States and Europe in many circumstances.

This perspective on housing and social control also throws some light on some of the issues raised by Professor Jackson. Partly because of suburbanism and the way the FHA and other federal housing programs worked, the urban poor no longer own their own homes to any great extent, and this may help to explain why the cities erupted in the 1960s. What is intriguing is that the skyrocketing price of housing is putting homeownership out of reach of more and more people in the middle and working classes, as well as those in the really impoverished, chronically unemployed class. So far, young families in the middle and working classes have managed to continue buying homes by devoting increasingly large proportions of their incomes to housing by sending both spouses to work, by deferring children, and by getting loans and other aids from parents. But it is not clear whether this will be a long-term solution to the continuing rise in prices. In light of this and the no-growth policies of many communities, it may be that homeownership rates will have to decline. If homeownership really is an important source of social stability in the United States, this, of course, may possibly put the future stability of the country in doubt.

Peter A. Marcuse:

Both of these chapters make important contributions to the understanding of housing policy and to why the particular configurations of housing location (in one case) and governmental support (in the other) came to be. Niethammer's essay is an exciting effort to go underneath the details of particular housing policies to show how in its broad sweep housing policy fit into the efforts of rising capitalism to keep under control new social forces in the cities it had itself unleashed. Jackson's essay proves conclusively what many students of housing have long suspected: the extent of direct governmental involvement in those urban processes that resulted in middle-class suburbanization and central-city ghettoization in the United States since 1930.

Both essays reject simplistic "benevolent state" explanations for why government has acted as it did (although Jackson seems to consider the alternative theory of conspiracy, which it is not). Both provide solid evidence to substantiate their arguments. Ken Jackson from original documentation not hitherto available, Lutz Niethammer by references not generally familiar here and certainly not hitherto used in the broad theoretical context in which he places them.

Yet neither essay is, I am afraid, ultimately satisfying in the proof of the proposition it sets forth. Let me first deal with what I see to be the major flaw in each chapter, and then turn to a consideration of some of the important substantive contributions that both nevertheless make. I deal with the flaws with some reluctance, not only because both essays are so valuable, but also because in fact I happen to believe that the basic argument each makes is absolutely correct.

Ken Jackson's contribution deals with the questions of the relationship between deliberate government policy and "unconscious" or "natural" market forces, consumer preferences, private prejudice, and so on, the process of suburbanization and differentiation between central city and suburb. He initially phrases this topic much more broadly:

> Whether the results of those [governmental] policies were foreseen by a government anxious to use its power and resources for   social control of ethnic and racial minorities.

But in fact he does not deal with the issue of social control in the body of the chapter, and the question of intent—whether there were specific government officials whose desire was to achieve precisely those long-term outcomes of their policies that did, in fact, ensue—is also left hanging. The question of intent, as any lawyer knows, is a tricky one at best, and often comes down to laying out the evidence about what inevitably *was* the outcome of a given act, leaving it to the reader (or the jury) to decide whether any reasonable person could have intended anything else in performing the act. I doubt whether any better evidence is available than what Jackson has produced, and he used what he has very effectively.

The real question of fact to which Jackson's evidence is addressed, then, is whether the federal government led or followed broader private forces in the process of suburbanization. Actually, Jackson hedges his own conclusion:

> The FHA was not the sine qua non in the mushrooming of the suburbs.... But in a critical period...the national government...developed policies that had the result...of the practical abandonment of large sections of older, industrial cities.... This same result might have been achieved in the absence of all government intervention, but the simple fact is that the various federal policies toward housing have had essentially the same effect.

So the importance, the relative contribution to the result—suburbanization and ghettoization—of federal policies and private forces is left open. But at least on the sequence of events, Jackson does take a position.

Washington actions were later picked up by private interests. This, I take it, is the hard core of his argument, and it is the evidence for this statement that I want to examine. There seem to me to be two problems with it. First, *why* did "Washington" act as it did? Was it acting out of its own initiative, because it concluded, through a process of independent reflection and deliberation, that these policies were "best"? Jackson himself provides evidence to the contrary: Social scientists had suggested certain trends (i.e., "invasion" of new groups—read minorities—into older, previously homogeneous areas) to be matters of fact, which appraisers read, again presumably as matters of fact, as having a clear influence on property values. Realtors and bankers, presumably having their enlightened self-interest in mind, cooperated with HOLC in defining and grading areas for mortgage-lending purposes. Could one not as easily argue that HOLC was reflecting the interests of private mortgage lenders and property owners in its policies, rather than leading or steering them? In his conclusion Jackson says, "suburbanization was an ideal government policy because it met the needs of both citizens and business interests." Which produced which—the policy the interests or the interests the policy—may then be debated.

I take it the question of causation or explanation is occasionally of some concern to historians, and that one needs at least to know why the question is being asked to know how to answer it. If the question here is, did the federal government and its agencies play a leading role in producing the present spatial patterns of U.S. cities, Jackson's evidence is overwhelming that it did. But if the question is, were these policies initiated by governmental actors and followed by private interests, or did government in turn act as it did because private interests were thus best served, the answer may be quite different.

Second, the data Jackson produces seem to me to need some further analysis before the conclusions he draws from them can be justified. Look at Table 4.1, for instance. From it the conclusion is drawn that, despite the invidious character of the rating system, HOLC's actual loan policies were not affected by it; unlike FHA, HOLC is exonerated from the charge of discrimination in deed, if not in word. But can one really tell, from the percentages in Table 4.1? More than 28 percent of HOLC's loans were made in hazardous areas, it is true; but what percentage of the need was there? HOLC, after all, was in the business of refinancing loans in default or threat of foreclosure; perhaps very few homes in A neighborhoods were in that situation. Perhaps, indeed, there were simply many less homes with mortgages in A neighborhoods. Don't we really need to know before we can judge? The implication of Table 4.1 is that one might expect an equal percentage of loans in each of the four types of neighborhoods, and the departures from equality are significant per se. I am not sure that implication is valid.

(Nor, parenthetically, can the last column of Table 4.4 be used, it seems to me, as quantitative evidence of discrimination by FHA. What, after all, was the level of single-family ownership, of new construction, or real-estate activity in the Bronx compared to that in Fairfax County, Virginia, in the period in question? That the program design of the FHA structurally discriminated against areas of multifamily housing, of slower growth, and of lower income is true, and graphically demonstrated; but that bureaucratic administration of the program based on arbitrary redlining produced these results it would take more than the data in Table 4.4 to demonstrate—although it probably is, in fact, true, and the dramatic nature of the differences in Table 4.4 do suggest it.)

Lutz Niethammer's essay is far more theoretical than Jackson's, and relies much less on hard data to make its argument. It is a welcome, and I think largely successful, effort to place housing policy in "a wider framework of social change and control." It opposes what I have elsewhere called the myth of the benevolent state, not a conspiracy theory of history (which I would not espouse either), but rather a view of policy as contributing to social control in a broad and pervasive sense. The passage in which Niethammer speaks of the complexity of the perceptions of housing—the sense in which issues of housing seem to tie into a sense of a whole culture at risk—seems to me to express something one sees in zoning hearings, in fights against low-income housing, in cross-burnings and two-acre zoning, in this country. And the descriptions of the reformers' view of working-class life and habits in the mid-nineteenth century—decay, filth, cynicism, corruption, savagery—may be much more similar to many perceptions of black ghettos in the United States than the etiquette of public discourse today would permit us to concede. But, again, I think there are flaws in the argument to the extent it is presented in this chapter, again, I think there are two main ones that need to be dealt with.

The heart of Niethammer's argument he summarizes as follows:

The housing debate…may be conceptualized as the experimental formulation of a new paradigm of social control, combining culture and nature and offsetting the predominant mixture of politics and economics. The resulting program, however, was in general too expensive to be acceptable to the dominant private interests of rising capitalism and went through a series of dismantling experiments to make it less costly, until the rising lower-middle and upper working classes advocated it in their demand for social space, and political intervention effectuated it in a much modified form.

The evidence Niethammer presents is a series of striking quotations from writers and speakers of the period, interpreted and linked together in an entirely persuasive way to substantiate his argument. But we are not provided with any framework by which we can judge the role of these writers; they are selected, as he himself admits, "almost at random." We know that Chadwick, in England, and Hobrecht, in Germany, were indeed major figures in the housing reform movement; but is Le Play's role in France of similar importance in this field? And to what extent did these three represent the major emphases of the debate? What other elements were there to be considered? Obviously no quantitative answers can be given to such questions, but a little more of the entire picture needs to be painted before these three particular sources can be properly put in their place within it.

This leads me to the second, and more general, weakness in Niethammer's paper. He says "*the* housing debate may be conceptualized as *the* experimental formulation of a new paradigm of social control." Are there then no other housing debates? Are there no other formulations that enter into it? No economic, political interests? No parties with other concerns? No other stakes?

Even the quotations Niethammer provides us suggest some of the other concerns that did, in fact, play their part in the overall debate. Chadwick clearly has major concerns with two: the danger of epidemics even to those not directly in inadequate housing, and the need to provide for the physical strength of the laborers. Neither of these is really a matter of maintaining relationships of domination and power, or of avoiding social unrest. And Niethammer concedes that in some cases employer provision of housing did occur, and for other, but very practical, reasons: No housing existed where production could best take place, i.e., the Ruhr. Certainly these factors play a role too; the new paradigm of social control is one thread, but only one, in the ongoing debate. In terms perhaps more familiar in Europe than in the United States, if legitimation and accumulation needs together explain most of governmental action, then Niethammer has made a major contribution to elucidating the meaning of legitimation and the role seen for housing in it; issues of accumulation, however, are not really touched upon in his chapter.

Let me conclude with a question, one addressed to both chapters and one raised in common by both—one for which both chapters provide much food for thought, but one which, in the end, neither answers. The question is terribly simple: "Why?" Why did the event they recount happen? An unfair question, perhaps, but ultimately the most important one. In the HOLC/FHA case, why did government act as it did in the 1930s and 1940s, and why did it stop doing so in the 1960s and 1970s (if it did)? Jackson rejects conspiratorial theories (rightly, I think); his note 119 suggests some congruence with interest-group theories; his conclusion seems to rely on cultural explanations ("the mainstream of public opinion," "America's affinity for the single-family house," "the preferences of a majority"). Which, or what combination, would he choose?

Lutz Niethammer, dealing directly as he does with the causes of specific positions on housing policy, moves the question one step back: We are told (I think persuasively) that much of housing policy can be explained by the attempt to impose a new "paradigm of social control"; but who is making that attempt, and why does it succeed or fail? Much of his explanation seems consistent with some current Marxist theory, yet he explicitly rejects "the dominant interests of capital and labor" as adequate to account for what he describes, and he views the reformers, the bearers of the new paradigm, as "outsiders to their ruling class" and "prophets in the desert." How, then, explain the developments he describes?

Why?

III
THE ECONOMY OF CITIES

The essays in this section by Brian Berry and Stephan Jonas attempt to look to the urban future in light of the past. However, the contributors are different in tone, language, attitudes, and their use of history.

Berry, while recognizing the importance of government policy, suggests that private market forces play a very important role in his consideration of inner-city futures. The geographer and planner employs historical analysis and data to deal with a contemporary problem: the fate of the nation's inner cities and the extent of the impact of "gentrification," the term used to describe the return of those who fled, or more likely whose parents fled, America's cities after World War II. Recent attention has focused on this obverse of "white flight." Observers, often with little historical insight, have seen a major trend building in America's cities as young professionals, frequently unmarried, or couples without children, restore brownstones, shop at trendy boutiques, and dine in fashionable restaurants. However, the 1980 census shows that older industrial cities continue to lose population despite the much-touted back-to-the-city movement. Even when metropolitan areas rather than cities are measured, the census indicates that of the 39 largest SMSAs in 1980, 7 lost population. These were New York, Philadelphia, St. Louis, Pittsburgh, Cleveland, Milwaukee, and Buffalo. In metropolitan areas that did grow, the core cities often lost residents; for example, Detroit lost 21 percent of its city population during the 1970s but gained 3 percent in its metropolitan area. Cities that do attract new residents—mostly young, single people or childless couples—continue to lose families with children. In fact, for the first time since 1820, rural areas and small towns by 1980 grew at a faster pace than the nation's metropolitan areas.

Studies have shown that this movement away from cities is based on the relocation of industries, businesses, services, and educational institutions to once remote areas. It is also premised on increasing ease of long-distance commuting via expressways, the growth of retirement and recreational communities in rural areas, and the renewal of mining. Hence, the 1980 census and similar findings are in agreement with Berry's analysis.

To understand inner-city decline, Berry contends, one must comprehend the dynamics of the U.S. housing market. The historical record of urban expansion in the twentieth century follows the peaks and troughs in the rate of capital formation in the housing sector. The link between new housing construction and inner-city depopulation is straightforward for Berry. Replacement housing construction—the housing built in excess of household growth—determines abandonment in the inner city, which he defines as the area substantially constructed prior to the Great Depression. The nation's housing policy has been de facto the only explicit national policy for urban development, and it has led to far-flung metropolitan regions and the transfer of growth to the least urbanized areas. Developments in transportation and communications, according to Berry, have diminished the need for large, dense urban areas. He expects the dilemma of the inner city to continue and, perhaps, to deepen as a result of the general aging of the nation's population, which will shift housing choices away from that of the trendy young, the continued decentralization of postindustrial employment, and the decline in geographic mobility.

Berry notes that a continued recession that cuts deeply into the new housing industry, or a crisis atmosphere, which brings forth "enlightened leadership," might turn things around and permit substantial inner-city revitalization. He does not, however, anticipate the lessening of the historic trend toward decentralization. His analysis, moreover, avoid directly grappling with the high cost of energy and skyrocketing mortgages, as several conference participants were quick to point out. As mentioned, however, the 1980 census revealed that these factors were not having an overall centralizing effect. If such factors actually were to force inner-city revitalization on a large scale, then the issue of those displaced from upgraded neighborhoods becomes paramount. With the poor pushed out to the suburban fringe, especially the inner suburbs, then urban historians will be acutely aware of a return to patterns, which in America were preindustrial. At the meeting, John Sharpless predicted the 1980s to be the era of the suburban slum.

Seymour Mandelbaum, also commenting on Berry's work, raises the issue of whether Berry's broad-scale consideration of secular trends might have been more finely honed with an analysis that would reveal small but important policy decisions. That is, do cultural preferences and inevitable aging shape changing neighborhoods, or is their history molded by political decisions that were and continue to be vulnerable to change? Such change is not limited to the American city, as David Goldfield, who is studying the renewing of urban Europe, has recently pointed out. While most of urban Europe's ancient historic centers are being preserved, the not-so-historic districts near the old centers are suffering. These working-class areas, erected during the late nineteenth and early twentieth centuries, suffer from the "American disease."

It is just this "Americanization" of the European city—decentralization leading to suburbanization and urban sprawl—that leads Stephan Jonas to call for a reexamination of European urban policy, especially in light of the decrease in economic growth since the middle of the 1970s. He discusses two approaches. One, French in origin, is a neoliberal, pre-Mitterrand "soft" doctrine of urbanization, which accepted growth and emphasized planning and the maintenance of centralized government power to protect the environment and to slow down such growth. The other is an alternative doctrine, originating with the Italian left, which emphasizes increasing urban unity within existing boundaries. It aims to reorganize the city rather than to enlarge it.

Mandelbaum, however, questions Jonas's initial premise that U.S. urban growth—"the American disease"—which may have served as a model for post World War II European cities, was ever "wild" or "unregulated." It may have been expansive and cities may have changed, but by no means did they dissolve. Moreover, the commentator contends that if Jonas wishes to deal adequately with the issues suggested by his urban alternatives, the sociologist must more effectively employ his historical arguments. Coping with the urban future, then, may well be facilitated by clearly understanding the urban past.

7

INNER-CITY FUTURES

An American Dilemma Revisited

BRIAN J. L. BERRY

A NEW AMERICAN DILEMMA

Gunnar Myrdal posted two interrelated dilemmas, the first of race relations and the second of differential regional growth caused by what he called a process of circular and cumulative causation. He saw the play of forces in the market as tending to increase rather than decrease regional inequalities. This, he said, was the consequence of the clustering of activities in areas that promote increasing returns, a result of the internal and external economies that are present in centers of agglomeration. He believed that the agglomeration advantages of the major northeastern urban-industrial complexes so swamped the cheaper factory prices of the periphery that they produced a continuous stream of disequilibriating flows of labor and capital from poor to rich regions. He concluded that free trade in an interregional system will always work to the disadvantage of poor regions, inhibiting their growth prospects and distorting their pattern of production. This led him to strong advocacy of governmental intervention to correct what

Editor's Note: *Reprinted from* Transactions of the Institute of British Geographers, *Volume 5, Number 1, 1980, with permission of the Institute.*

he perceived to be the "normal tendencies" in a capitalist system to sustain and increase inequality.

The kind of central government controls sought by Myrdal never came. Instead, the key premise underlying policy development in the United States remains the utilitarian belief that solutions to the nation's needs must be found, for the most part, in the private sector. Market processes are relied upon to allocate resources efficiently and to provide new jobs, rising incomes, and better housing. An essential prerequisite to realizing differential market opportunities is believed to be sufficient mobility of capital and labor; indeed, the more the better. The principal roles of government are, thus, those of regulator, facilitator, and occasionally social engineer in preserving, supporting, and enhancing mainstream objectives: providing information if it is lacking on the part of buyers or sellers; preventing emergence of undue concentration of economic power which results in higher prices and fewer services than if competition prevailed; reducing market fluc-tuations and cushioning their consequences; and facilitating mobility—in other words, promoting the mainstream values of democratic pluralism. Other forms of government intervention are believed to be justified only if public welfare is endangered and if adequate remedies are not available in the marketplace. There are several cases. For example, if market prices do not reflect the full social costs or benefits of development because of congestion or external costs such as pollution, noise, and other man-made hazards, too much or too little of a good or a service will be provided, unless the competing interests of those who benefit and those who pay are arbitrated. Similarly, if there is an inability to determine or collect a proper price, as in the case of a public good whose consumption by one individual does not reduce the consumption of it by others, misallocations will result if the situation goes uncorrected. A third case arises if there are demonstrable advantages to society from main-taining minimum levels of service to population groups or communities that otherwise would be unable to obtain it. Frequently, such minimum levels of service are characterized as basic "rights."

As late as 1970, U.S. urban policy was being formulated in the belief that the circular and cumulative causation described by Myrdal would continue, and that increasing urbanization did, in fact, endanger the public welfare in a variety of ways, demanding reactions that would both redirect mainstream growth in the future and correct the problems resulting from cumulative growth in the past. Thus, Title VII of the Housing and Urban Development Act of 1970 (Public Law 91-609, 84 Stat. 1791; 42 U.S.C. 4501) stated in Section 702 that

> the rapid growth of urban population and uneven expansion of urban development in the United States, together with a decline in farm population, slower growth in rural areas, and migration to the cities, has created an imbalance between the Nation's needs and resources and seriously threatens our physical environment.... The economic and social development of the Nation, the proper conservation of our natural resources, and the achievement of satisfactory living standards depend upon the sound, orderly, and more balanced development of all areas of the Nation.... The Congress...declares that the national urban growth policy should...favor patterns of urbanization and economic development and stabilization which offer a range of alternative locations...help reverse trends of migration and physical growth...treat comprehensively the problems of poverty and employment...associated with disorderly urbanization and rural decline.

To those who drafted the 1970 Housing Act, Myrdal's American dilemma was a continuing reality.

Barely a decade later, beliefs about the nature of change appear to have reversed, however, and along with them the apparent growth directions to be promoted and the corrective reactions to be taken. Thus, President Carter's Urban and Regional Policy Group's March 1978 report, *A New Partnership to Conserve America's Communities: A National Urban Policy*, declared as follows:

> Three major patterns of population change can be traced in the Nation today: migration from the northeastern and north central regions of the country to the south and west; the slower growth of metropolitan areas and the movement from them to

small towns and rural areas; and movement from central cities to suburbs.... Today's widespread population loss in the Nation's central cities is unprecedented.... The thinning out process has left many people and places with severe economic and social problems, and without the resources to deal with them.... Our policies must reflect a balanced concern for people and places...to achieve several broad goals: [to] preserve the heritage and values of our older cities; maintain the investment in our older cities and their neighborhoods; assist newer cities in confronting the challenges of growth and pockets of poverty... and provide improved housing, job opportunities and community services to the urban poor, minorities, and women.... If the Administration is to help cities revitalize neighborhoods, eliminate sprawl, support the return of the middle class to central cities, and improve the housing conditions of the urban poor, it must increase the production of new housing and rehabilitation of existing housing for middle class groups in cities....We should favor proposals supporting: (1) compact community development over scattered, fragmented development; and (2) revitalization over new development.

The new American dilemma, at least in the eyes of those who would remake U.S. urban and regional policy, is apparently the reverse of Myrdal's. The issue has become that of the older inner cities, the former leaders in the nation's growth, now believed to be suffering not from increasing size and density, but from employment declines, population losses, ghettoization of minorities and the poor, physical deterioration, and fiscal distress.

BACKGROUND: REGIONAL SHIFTS
AND DEMOGRAPHIC CHANGES

That major changes have taken place and that many of these changes involve a reversal of Myrdal's model is without question. Contrary to Myrdal's expectations, regional incomes have converged since 1929, not diverged (Figure 7.1). Nominal income covergence is only part of the story, however. Some of the nation's most affluent urban-industrial regions in 1929, New England in particular, slipped to the bottom of the ranking in real income terms by 1977 (Table 7.1).

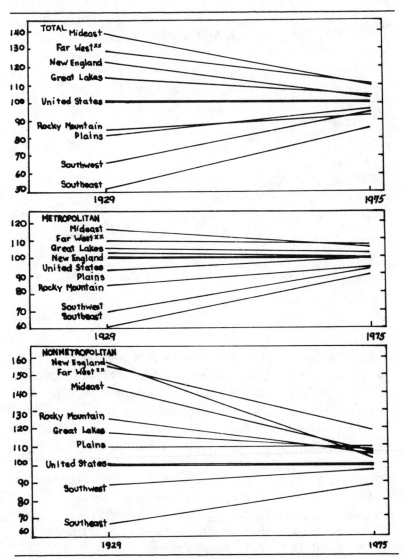

Figure 7.1: Convergence of Nominal Regional Per-Capita Income in the United States, 1929-1975

SOURCE: Daniel H. Garnick, "A Reappraisal of the Outlook for Northern States and Cities in the context of U.S. Economic History." Joint Center for Urban Studies, Cambridge, Mass., Jan. 1978, based on data from U.S. Department of Commerce, Bureau of Economic Analysis.

NOTE: Far West includes Alaska and Hawaii. Per-capita personal income is total personel income divided by population as of July 1.

TABLE 7.1   Regional Per-Capita Incomes, 1977, (indexed): Nominal
             and Adjusted for Cost of Living

Region	Nominal	Adjusted
New England	102	88
Mideast	107	98
Great Lakes	105	103
Plains	97	98
Southeast	86	93
Southwest	95	101
Rocky Mountain	94	98
Far West	111	106

SOURCE: U.S. Department of Commerce, Bureau of Economic Analysis, *Survey of Current Business,* August and October 1978. Cost-of-Living adjustment based on Frederich J. Grasberger, "Developing Tools to Improve Federal Grant-in-Aid Formulas" (Formula Evaluation Project, Preliminary Report #3, Center for Governmental Research, Rochester, New York, 1978). Comparison adapted from Advisory Commission on Intergovernmental Relations, *Income Growth Differential Study* (1978) by George E, Peterson and Thomas Muller of The Urban Institute, Washington, D.C.

These regional shifts are in large measure a result of the changing location of American industry, the breakdown of the traditional heartland-hinterland organization of the U.S. economy in which leadership in technological innovation and industrial growth was exercised by the major urban centers of the northeastern manufacturing belt. For the first half of the twentieth century, the manufacturing belt accounted for some 70 percent of the nation's industrial employment. Between 1950 and the mid-1960s, manufacturing jobs continued to grow in the Northeast, but the growth was more rapid in other regions of the country and the manufacturing belt's relative share fell to 56 percent. By 1970 relative decline had been replaced by absolute losses. From 1969 to 1977 the manufacturing belt lost 1.7 million industrial jobs, almost exactly the job growth of the former periphery. Relative growth of the service industries has paralleled these industrial shifts, multiplying the effects of the changes.

Traditionally, the manufacturing belt was the center of innovation. It was able to introduce new industries to offset losses of standardized industries to cheap-labor areas elsewhere. But this is no longer the case. The economy's rapid growth industries (electronics, aerospace, scientific instruments, and so on) are dispersed throughout the former periphery; it is the older slow-growth industries that remain in the former core. Employment in these remaining industries is extremely sensitive to cyclical change in the economy, which compounds the distress of northeastern cities when the economy is in recession. But what is even more critical is that the manufacturing belt appears to have lost its traditional seedbed function. The locus of innovation and growth has shifted elsewhere.

Job shifts have been accompanied by population shifts. Following the bulge in the population pyramid formed by the post-World War II baby boom, there has been a decline in fertility rates to less than replacement levels. As natural increase has diminished, migration has become an increasingly important source of population change. This growing importance of migration as a factor of growth has been intensified by the movement of the baby boom cohort into its most mobile years. In all urban-industrial countries, a certain minimum amount of geographical mobility is a structured part of the life cycle, with the greatest rates associated with the stage at which young adults leave the parental home and establish an independent household shortly after formal schooling is completed. The baby boom cohort is now passing through this stage, and the subsequent period in which spatial differences in real wage rates and in employment opportunities provide signals that encourage economically motivated migration. This migration not only increases the well-being of the movers themselves, but also results in improved resource allocation. Thus, job shifts in a period of maximum potential mobility have resulted in massive reversals of the migration streams described by Myrdal. Net migration has increased from manufacturing belt to periphery, for both majority and minority members of the U.S. population. The South has experienced a dramatic and accelerating migration reversal (Figure 7.2). Within regions, net migration reveals moves from central cities to suburbs and exurbs and from metropolitan to

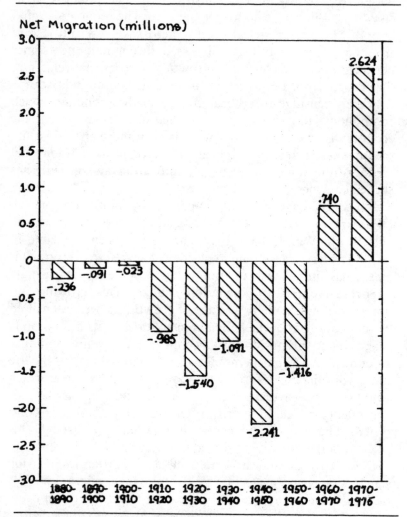

Figure 7.2: The South's Net Migration from 1880 to 1975

SOURCE: Larry H. Long, Interregional Migration of the Poor. Current Population Reports, Special Studies Series P-23, No. 73, Washington, D.C.: U.S. Department of Commerce, Bureau of the Census, 1978, p. 4. Reproduced with the author's permission.

nonmetropolitan areas (Table 7.2). Throughout the nation, migrating workers have left jobs located in major metropolitan cores for workplaces in smaller urban areas, suburbs, and nonmetropolitan America. Since 1970, the Northeast as a

TABLE 7.2 Migration Streams, 1970-1975, and Net Migration by Type of Residence and Race, 1970-77, Excluding Net Immigration from Abroad

Migration Streams 1970-1975 (in millions)			
Residential Category in 1970	Residential Category in 1975		
	Central City	Suburb	Nonmetropolitan Area
Central City	[17.11]	9.8	3.2
Suburb	3.8	[18.2]	3.5
Nonmetropolitan Area	2.1	3.0	[19.0]

NOTE: Figures in brackets are movers who remained in same residential category during relocation.

Net Migration 1970-1977 (in thousands)			
Type of residence	All races	White	Black
Metropolitan Areas.....	– 2,260	–2,411	145
Central cities........	–10,451	–9,533	–653
Suburban areas......	8,190	7,122	798
Nonmetropolitan areas..	2,260	2,411	–145

SOURCES: U.S. Bureau of the Census (1975), "Mobility of the Population of the United States: March 1970 to March 1975," *Current Population Report* Series P-26, No. 285, 1976, Washington, D.C.: U.S. Department of Commerce; and *idem*. "Social and Economic Characteristics of the Metropolitan and Nonmetropolitan Population: 1977 and 1970," *Current Population Reports,* Series P-23, No. 75, 1978.

whole has lost population, a result of decreasing natural increase and of the net migration reversal; in the South continued high levels of growth have occurred, despite declining natural increase, because of increasing inmigration (Figure 7.3).

As a result, nonmetropolitan areas are now growing more rapidly than metropolitan areas, and central cities are declining, especially within the largest metropolitan regions (Table 7.3). Thirty of the nation's fifty largest cities have lost population since 1970; one in five registered a loss of at least 10 percent between 1970 and 1975. Because the incomes of out-migrants were greater than those of inmigrants, the income loss

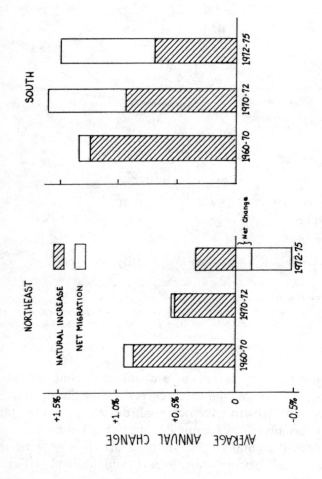

Figure 7.3: Population Changes in Northeast and South Before and After 1970

SOURCE: Peter Morrison, the Rand Corporation, with permission.

TABLE 7.3 Population by Type of Residence, 1977 and 1970 (numbers in thousands, 1970 metropolitan area definition)

Type of residence	1977	1970	Numerical change 1970 to 1977	Percentage change 1970 to 1977	Percentage distribution 1977	1970
United States	212,566	199,819	12,747	6.4	100.0	100.0
Metropolitan areas	143,107	137,058	6,049	4.4	67.3	68.6
Central cities	59,993	62,876	-2,883	-4.6	28.2	31.5
Suburban areas	83,114	74,182	8,932	12.0	39.1	37.1
Metropolitan areas of 1 million or more	82,367	79,489	2,878	3.6	38.7	39.8
Central cities	31,898	34,322	-2,424	-7.1	15.0	17.2
Suburban areas	50,469	45,166	5,303	11.7	23.7	22.6
Metropolitan areas of less than 1 million	60,739	57,570	3,169	5.5	28.6	28.8
Central cities	28,095	28,554	-459	-1.6	13.2	14.3
Suburban areas	32,644	29,016	3,628	12.5	15.4	14.5
Nonmetropolitan areas	69,459	62,761	6,698	10.7	32.7	31.4
Counties designated metropolitan since 1970	9,980	8,373	1,607	19.2	4.7	4.2
Other nonmetropolitan counties	59,479	54,388	5,091	9.4	28.0	27.2

SOURCE: U.S. Bureau of the Census, "Social and Economic Characteristics of the Metropolitan and Nonmetropolitan Population: 1977 and 1970." *Current Population Reports*, Series P-23, No. 75, Washington, D.C.: U.S. Department of Commerce, 1978.

of metropolitan areas (gain of nonmetropolitan areas) was over $17 billion between 1975 and 1977 alone.

THE SHAPING EFFECT OF
HOUSING MARKET DYNAMICS

Understanding why an inner-city problem should have emerged within the framework of these shifts—why the nation's leading central cities should now be the places where population declines are concentrated—demands that one understand the shaping effects of U.S. housing market dynamics. First, urban growth has been strongly cyclical in both the long and the short terms. American cities have not grown in a smooth or continuous manner, but in a series of major bursts, each of which has added a new ring of structures dominated by a particular building style. Nationwide, the magnitude of housing investment probably has done more than any other factor to shape urban growth. The historical record of urban expansion in this century closely follows the peaks and troughs in the rate of capital formation in the housing sector. From 1910 to 1914 and again from 1921 to 1929, when real-estate investment boomed, metropolitan boundaries surged outward; later, when housing investment nearly came to a halt during the Great Depression and World War II, urban expansion slowed to a virtual standstill; then, in the 1950s, an unprecedented volume of housing investment was accompanied by a record pace of suburbanization.

Regionally, the expression of these cycles is to be seen in the different size of successive housing stock increments. Cities in older growth regions, especially those in the Northeast, have several growth rings with substantially differentiated housing stocks; indeed, a good working definition of the "inner city" is that area substantially constructed before the Great Depression. Cities located in newer growth regions, especially those of the South and the West are dominated by housing built since World War II (Table 7.4).

Until World War II, less than half of the nation's population owned their own homes, and less than half of the housing stock were in single-family units. The great surge in homeownership

TABLE 7.4   Age of the 1970 Housing Stock in Three Metropolitan Areas

	Reading, PA		Chicago, IL		San Diego, CA	
	Total[a]	Single-Family[b]	Total[a]	Single-Family[b]	Total[a]	Single-Family[b]
Before 1880	5.4	19.9	31.0	16.3	0.5	31.3
1880-1889	4.6	10.7	77.9	12.3	1.4	24.1
1890-1899	8.1	11.8	175.4	14.6	2.7	38.1
1900-1909	9.9	13.0	223.2	17.7	7.1	55.9
1910-1919	5.7	31.7	252.2	23.4	17.6	66.9
1920-1929	8.7	40.1	416.9	32.7	32.1	80.1
1930-1939	2.5	65.0	62.3	64.0	21.4	81.7
1940-1949	5.4	75.3	165.3	77.0	68.8	65.5
1950-1959	15.1	49.0	488.0	75.0	154.0	70.1
1960-1969	15.5	65.0	492.3	55.5	150.0	57.5

SOURCE: U.S. Bureau of the Census.
a. In thousands.
b. Single-family residences as a percentage of the total built in the time period.

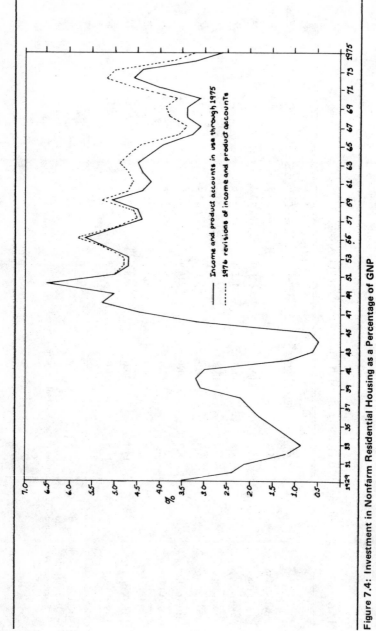

Figure 7.4: Investment in Nonfarm Residential Housing as a Percentage of GNP

SOURCE: Redrafted, with permission, from an original in George E. Peterson, "Federal Tax Policy and the Shaping of the Urban Environment," National Bureau of Economic Research, 1977.

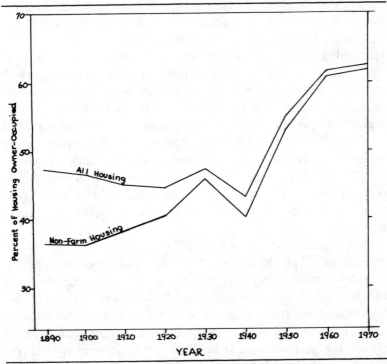

Figure 7.5: Trends in Homeownership in the United States, 1890-1970
SOURCE: Peterson, op. cit. With the author's permission.

came in scarcely more than a decade between 1948 and 1960 (Figures 7.4 and 7.5). This boom in homeownership followed hard on the heels of the effective introduction of tax subsidies for owner occupancy, a by-product of the mass income tax first adopted during World War II, and the formulation in the early 1930s of a national housing policy that sought to promote homeownership as a stabilizing social force relying heavily on new construction as the tool to upgrade national housing standards and to provide for needed geographic mobility of urban households seeking better neighborhoods.

In the United States, tax incentives are perhaps the chief instrument the government possesses for allocating investment resources among competing sectors of the economy. Housing investment reached its height during the years 1948 to 1960, partly because of the many new tax laws passed during or after World War II which singled out housing for favorable tax treatment. Historically, perhaps the most consistent bias in the federal tax code has been the favoritism given to investment in new structures relative to investment in the improvement and repair of existing structures, which accelerates the rate at which buildings are replaced. Although this speeding up of the replacement cycle for structure does not in itself give a locational bias to development, it compresses the period during which urban regions adjust to changed private market prices or new transportation technologies. When favoritism toward new construction is combined with other tax policies that favor homeownership, important locational effects do result, however.

It is a peculiarity of the tax subsidy method of investment stimulation that the value of the tax advantage granted to homeownership is proportional to an investor's marginal tax bracket, for those who claim the homeownership deduction. This deduction took on an allocative significance for the first time during World War II, when the marginal federal tax paid by most Americans rose from 4 percent to 25 percent, making the deductibility of homeowner expenses far more valuable than they previously had been. Today, the value of the tax benefit jumps sharply by household income level. At upper-middle income levels, the tax system creates an annual cost advantage for owner occupancy of about 14 to 15 percent.

The growth in homeownership rates shown in Figure 7.5 cannot be attributed solely to federal tax policy, of course. A number of other national policies—such as the introduction of FHA financing in the early 1930s, VA financing after World War II, and the opening up of the suburbs through highway construction in the 1950s—lent force to the homeownership boom. Household income growth also contributed to higher rates of owner occupancy because homeownership rates are a function of income level (Figure 7.6). Together, both directly and indirectly, these forces encouraged low-density single-family living patterns, with generous amounts of land consumption on the urban fringe, since large-scale single-family subdivisions require extensive land parcels that can be

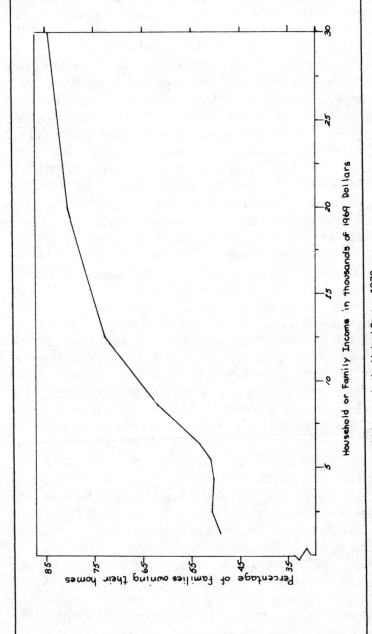

Figure 7.6: Homeownership and Household Income in the United States, 1970

SOURCE: Peterson, op. cit. With the author's permission.

203

assembled most easily at the urban periphery. On average, over the period 1950-1970, each newly constructed single-family home added approximately six-tenths of an acre to the nation's urbanized area, as defined by the Bureau of the Census. This is far more than the immediate lot space occupied by new housing. Although records on land usage in new construction are incomplete, data assembled by the National Association of Homebuilders indicate that the average lot size for new single-family homes in metropolitan areas was about .35 acre in the mid- to late 1960s. The effect has been that single-family construction has tended to stretch metropolitan boundaries.

The link between new housing construction and inner-city depopulation is straightforward. Since the early 1960s, new housing construction has far exceeded household growth (Table 7.5). Between 1963 and 1976 household expansion was some 17 million, but 27 million new housing units were constructed. The crucial role of housing construction in excess of household growth, which can be termed *replacement housing construction*, in determining the value and maintenance of older housing is not surprising. When new housing is built and occupied, more often than not by relatively well-to-do families, the older housing vacated by these families filters down to provide better-quality homes for lower-income families. It is the least desirable housing that is left vacant after this housing game of musical chairs comes to an end, ultimately to be abandoned, demolished, and perhaps to be redeveloped. More often than not, this housing is the oldest and the most outmoded remnant of the earliest housing cycles.

Preceding abandonment, housing maintenance falls with housing values, and the result is deterioration, both of older houses and of older neighborhoods. Thus, periods of abundant housing investment are years of vigorous new housing construction, and in the United States most new housing always has been built near the the urban periphery. But there is a natural tendency of a highly fragmented and speculative building industry to overshoot. Since 1960 more than one-third of all new housing construction has replaced older stock rather than adding to the total supply of housing. As a consequence, the high rate of housing production not only pushed the urban

TABLE 7.5   National Trends in Household Growth, New Housing Production, and Replacement Housing Construction, 1963-1975

Year	Numbers of Households		Numbers of Newly Constructed Housing Units	Replacement Housing Construction	
	Numbers	*Change from Previous Year (1000s)*		*Numbers (1000s)*	*Percentage of Households*
1963	55,189	—	1,786	—	—
1964	55,996	307	1,752	945	1.7
1965	57,251	1,255	1,727	472	0.8
1966	58,092	841	1,413	572	1.0
1967	58,845	753	1,562	809	1.4
1968	60,446	1,601	1,863	262	0.4
1969	61,805	1,359	1,913	554	0.9
1970	63,401	1,596	1,870	274	0.4
1971	64,778	1,377	2,582	1,205	1.9
1972	66,676	1,898	2,955	1,057	1.6
1973	68,251	1,575	2,625	1,050	1.5
1974	69,859	1,608	1,682	74	0.1
1975	71,120	1,261	1,384	123	0.2
1976	72,867	1,747	1,798	51	0.1

SOURCE: U.S. Bureau of the Census, "Household and Family Characteristics," *Current Population Reports*, P-20. After Franklin J. James, "Private Reinvestment in Older Housing and Older Neighborhoods," Committee on Banking, Housing and Urban Affairs, U.S. Senate, 1977; and U.S. Department of Housing and Urban Development, *Housing in the Seventies: A Report of the National Housing Policy Review* (Washington, D.C.: Government Printing Office, 1974) Table 2, Chapter 4.

NOTE: Numbers of households are measured in March of each year.

boundary outward, but also produced faster abandonment of older housing located in the inner city.

The critical variable in understanding abandonment in any particular housing market therefore is the magnitude of the replacement supply, which depends in turn on the extent to which new housing starts outpace the rate of household growth. A declining metropolitan area with a moderate rate of new housing starts (such as Cleveland or New York), according to this relationship, might be expected to evidence substantial abandonment of older inner-city neighborhoods, whereas a rapidly growing metropolitan region with massive new housing construction would not if household growth kept pace with housing starts; this is indeed the nationwide story. Net migration from manufacturing belt to sunbelt, from larger to smaller cities, from metropolitan areas, and from inner cities to the urban periphery, have concentrated abandonment in the older inner sections of the manufacturing-belt metropolitan areas.

Seen in this light, the nation's housing policy has been, de facto, the only explicit national policy for urban development that we have had. This policy has been direct and successful, promoting household wealth through homeownership, improved living conditions via new construction, and increased efficiency by means of mobility. Its very success as a facilitating force has contributed both to the emergence of a new scale of low-slung, far-flung metropolitan regions and to a new force of counterurbanization—the transfer of the locus of new growth to some of the most remote and least urbanized parts of the country. The settings where this growth is now occurring are exceedingly diverse. They include regions oriented to recreation in northern New England, the Rocky Mountains, and the upper Great Lakes; energy supply areas in the northern Great Plains and southern Appalachian coal fields; retirement communities in the Ozark-Ouachita uplands; small manufacturing towns throughout much of the South; and nonmetropolitan cities in every region whose economic fortunes are intertwined with state government or higher education. Factors contributing to these shifts appear to be: changes in transportation and communications that have removed many of the problems of access that previously served to constrain the growth prospects of the periphery, permitting decentralization of manufacturing on the inexpensive land and benefiting from the low wage rates of nonmetropolitan areas; the trend toward earlier retirement, which has lengthened the interval during later life when a person is no longer tied to a specific place by a job; and an increased orientation at all ages toward leisure activities, caused in party by rising per-capita income and centered on amenity-rich areas outside the daily range of metropolitan commuting. These are but symptoms of the more profound forces that are at work, however. The concentrated industrial metropolis only developed because proximity meant lower transportation and communication costs for those interdependent specialists who had to interact with each other frequently or intensively. One of the most important forces contributing to counterurbanization is therefore the erosion of centrality by time-space convergence. Virtually all technological developments of industrial times have had the effect of reducing the constraints of geographical space. Developments in transportation and communications have made it possible for each generation to live farther from activity centers, for these activity centers to disperse, and for information users to rely on information sources that are spatially more distant yet temporally more immediate. In other words, large, dense urban concentrations are no longer necessary for the classical urbanization economies to be present. Contemporary developments in communications are supplying better channels for transmitting information and improving the capacities of partners in social intercourse to transact their business at great distances and at great speed. The time-eliminating properties of long-distance communication and the space-spanning capacities of the new communication technologies are combining to concoct a solvent that has dissolved the agglomeration advantages of the industrial metropolis, creating what some now refer to as an urban civilization without cities. The edges of many of the nation's urban systems have now pushed one hundred miles and more from declining central cities. Today's urban systems appear to be multinodal, multiconnected social systems sharing in national growth and offering a variety of lifestyles in a variety of environments. And what is being abandoned are those environments that were key in the traditional metropolis-driven growth process: the high-density, congested, face-to-face center-city settings that are now perceived as aging, polluted, and crime-ridden, with declining services and employment bases and escalating taxes. Such is the new American dilemma.

WHAT PRICE THE INNER CITY?

Seen in this light, abandonment may be viewed as a measure of the success we have achieved in our housing policy and in responding to new growth opportunities. Yet the inner city cannot and should not be so lightly written off. There is clearly a national interest in the problems of the central city that derives from three distinguishable premises:

(1) That it is socially wasteful to underutilize, abandon, or destroy capital investments made by preceding generations in urban infrastructures, housing, places of business, and public buildings. The national product is diminished by our present course of reproducing these facilities elsewhere rather than using what already exists.

(2) That although their populations are decreasing, central cities still hold a very large number of people whose lives and fortunes are unfavorably affected by their physical and social environments. Even assuming that suburbanites and exurbanites entirely escape these adversities, over a fourth of the nation's people live in central cities and contend daily with their stresses. The welfare of these people is surely a matter of national interest.

(3) That although central cities, as large, dense concentrations of people and jobs, have become technologically obsolete, the shift to a new spatial organization can perhaps be made less painful to those who must adapt to it. Orderly change would be less costly to society as a whole than allowing stresses to accumulate within a system trying desperately to maintain itself until the system as a whole fails.

National concern about these problems has been evident in a series of federal programs designed to reverse or at least to slow down the decay. Over time, these programs have reflected a widening perception of the interplay of forces that cause central-city deterioration, have tested a variety of remedies, and have spent many billions of dollars, all to little avail.

The sequence of federal programs began in the 1930s with emergency public works and public housing construction, supplemented in 1949 by a large commitment of federal funds to slum clearance and redevelopment. Beginning in 1954, the emphasis shifted to less capital-intensive programs of housing rehabilitation and neighborhood preservation. Along with this changed emphasis came a growing array of capital grants to improve the urban infrastructure, and increasing support for the common functions of city government.

The urban riots and racial strife of the early 1960s testified that the urban poor did not believe in the efficacy of these remedies. President Johnson declared "war" on poverty, and his new programs stressed both neighborhood political mobilization and integrated planning to improve the physical, social, and economic circumstances of the urban poor and racial minorities. Dissuaded by the apparent ineffectiveness of these programs, the Nixon administration withdrew the federal presence from the cities, turning urban problems back to city hall. Instead of advice and supervision, Washington would dispense only money.

In each of these episodes, the Congress and the administration were responding partly to new circumstances, but mostly to the perception that the programs confidently offered a few years earlier were ineffective or even perverse in their consequences. It may be that the mistakes they recognized retrospectively form a larger pattern that we can at this greater distance perceive, for the post-1970 evidence indicates, as we have seen, that deterioration is spreading to more cities and accelerating in those where it has long been evident.

First, the majority of these urban programs failed because they were targeted toward areas left behind by the forces of change, forces that in turn were reinforced by the more effective de facto urban policy embodied in the nation's housing programs.

Turning the housing equation around, investment in new housing construction is to some extent a substitute for investment in the preservation, repair, and upgrading of old housing. High interest rates and high labor and material costs tend to discourage housing investment of all kinds, but the same price changes enhance the value of the existing housing stock by making it more expensive to reproduce. It then becomes more economical to preserve and improve buildings rather than to allow them to deteriorate. New housing starts increased rapidly during the building boom of the early 1970s and the numbers of building permits taken out in central cities for residential alterations and additions slumped. After 1974, as the housing industry entered a period of combined inflation and recession, new housing starts lagged and private-market revitalization of the inner city began to take hold. The fixed-interest-rate mortgage system in use in the United States provides a financial motive for this kind of upgrading of existing housing during periods of capital market tightness. Many homeowners would have to surrender their old, low-interest mortgages if they acquired new homes; such households often find it more profitable to satisfy their demand for more housing by making additions, replacements, or alterations in their present homes, while retaining their old mortgages. Considerations like these have made the home repair industry one of the strongest countercyclical sectors in the American economy. When total housing investment is high, rehabilitation and repair expenditures tend to lag; when total housing investment declines, capital expenditures on the standing stock actually intensify.

Unfortunately, both cheap, plentiful new housing and vibrant older neighborhoods in central cities are desirable goals. But if no choice is made between conservation and the broad-scale encouragement of new housing production and both goals are pursued simultaneously, the results are likely to be extremely wasteful. The national housing experience during the early 1970s illustrates the costly waste resulting from uncoordinated attempts to combine large-scale new housing production with large-scale conservation of older housing. As the 1960s drew to a close, the nation undertook major new initiatives to achieve the goal of a decent home in a decent neighborhood for everybody, the poor as well as the rich. Major new federal programs set up in the Housing Act of 1968 produced a massive upsurge in both federally subsidized new housing construction and in federally subsidized housing rehabilitation. By 1970 over 425,000 new units were receiving direct federal subsidies. This amounted to more than one in five new units being built within the nation. Federally subsidized housing rehabilitation also grew apace. In 1972, 51,000 units were rehabilitated using federal subsidies. Federal rehabilitation programs were focused largely on central-city homeowners and landlords. It appears that the federal government subsidized between 8 and 11 percent of the total dollar volume of reinvestment in owner-occupied housing in central cities during the early 1970s.

The enormous levels of new housing production encouraged by these subsidies during these years cut deeply into private demand for the existing stock of owner-occupied housing in cities. Private, unsubsidized reinvestment by central-city homeowners fell more rapidly than federally subsidized reinvestment increased. Between 1969 and 1970, federally subsidized expenditures for the rehabilitation of owner-occupied homes in cities are estimated to have doubled, increasing by $150 million (measured in constant 1973 dollars). In the same years, private unsubsidized reinvestment by central-city homeowners fell by $250 million, or 10 percent.

Thus, the attempt to combine massive new housing production with large-scale rehabilitation of the existing housing stock cut into housing values in central cities, encouraging mortgage defaults and abandonment, and thus contributed to the crisis in federal housing policy that brought on President Nixon's moratorium, putting an end to many of the housing programs that underlay the federal housing initiatives.

PRIVATE-MARKET INNER-CITY REHABILITATION: PROSPECTS AND LIMITS

Since the early 1970s, however, there have been signs of private-market renovation of some neighborhoods in some central cities, a function of both a slowdown of new housing starts and of lifestyle shifts in the baby boom generation. During the 1960s privately renovated neighborhoods such as Georgetown in Washington, D.C., Greenwich Village in New York, or Boston's South End were dismissed as unique. But more recently homeowners have begun to renovate old neighborhoods all across the country.

Part of the reason for revitalization is certainly the downturn in new housing starts since 1974. The metropolitan areas with the lowest rates of replacement supply have been those which, in general, have experienced the most rapid inflation of housing prices and the most substantial neighborhood revitalization; those with the greatest replacement supply have had the least revitalization. But there is another source of variation among metropolitan areas, on the demand side. Since 1970 there have been many changes affecting the number and nature of new households entering the housing market. An increase in owner-occupancy rates from 69 to 75 percent between 1970 and 1975 has been accompanied by rapid change in the nature of baby-boom-generation households.

The increase in homeownership, occurring most rapidly among one-person or single-parent households, took place simultaneously with the massive inflation of housing costs and surely represents an investment rather than a consumption decision: an inflation hedge against being priced out of the market in the future. The inflation, in turn, has been both cause and the effect of the slower pace of new housing starts which of itself should have made inner-city reinvestment more attractive.

But overlaid on this have been the housing preferences of new higher-income young homeowners not pressed by child rearing, with two workers, one or both of whom may be professional, seeking neighborhoods in the inner city with geographic clusters of housing structures capable of yielding high-quality services; a variety of public-good amenities within safe walking distance of these areas, such as a scenic waterfront, parks, museums or art galleries, universities, distinguished architecture, and historic landmarks of neighborhoods; and a range of high-quality retail facilities and services, including restaurants, theaters, and entertainment.

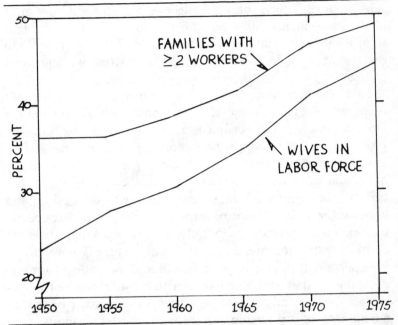

Figure 7.7: Families with Working Wives and Two or More Paychecks in the United States, 1950-1975

SOURCE: U.S. Census data.

These preferences follow directly from lifestyle and compositional shifts. First, continued development of American society has resulted in increased economic parity for women; this enables them to have the option of roles other than that of housewife and mother. In consequence, men and women lead more independent lives, and are able to exercise more options in life-course transitions. Increasing numbers of couples live together without the formal ties of marriage. The direct and opportunity costs of child rearing are rising, birth-control technology has improved and abortion laws have been liberalized, and hence the birthrate is dropping. There are increasing numbers of families with two or more workers, and there are more working wives than ever before (Figure 7.7).

Revitalization, then, has been taking hold first in superior neighborhoods in some of those metropolitan areas that have the lowest rates of replacement supply—in particular, the metropolitan areas that have a sizable cluster of professional jobs supporting the youthful, college-educated labor force most likely to evidence lifestyle shifts. An implication is that significant revitalization may be limited to the metropolitan centers with agglomerations of postindustrial management, control, and information-processing activities.

Table 7.6 highlights the relationship of type of housing unit occupied to stage in life cycle, showing a shift away from the free-standing, owner-occupied, single-family house between 1960 and 1970 for younger households. This is because smaller households require less space, because the fluidity of households and the looser legal links among their members is contrary to the rigidity of tenure associated with ownership, and because the maintenance of such a house and its grounds is time-consuming since, with most people working, time for domestic work becomes scarce and costly. There is a preference for apartments or row or town houses and innovative forms of design, as well as experimentation with forms of tenure (such as condominiums and cooperatives) that preserve some of the tax advantages of ownership but provide greater liquidity, and increases in new forms of contracting arrangements for the operation and maintenance of housing.

Increasing attractiveness of more central locations in the core city and the older suburbs follows, for it is there that there is an appropriate stock of housing and access to services as well as locational convenience for the journey to work. Since many of these households have no children, the racial factors of school integration do not act as they have in the white flight to the suburbs. The suburbs are thought of as good places in which to raise children. Obviously, this plays no part in the location preferences of those without children.

But there are polarizing effects here as well. Over the past twenty years the housing of the poor and the working poor has improved primarily because they have fallen heir to what used to be called "the gray areas." The softening of middle-class demand for this housing stock lowered its relative price and permitted a sharp decline in overcrowding for low-income people. Whatever the troubles of the cities, this has been a fortunate outcome. But the danger appears imminent that the

TABLE 7.6 Housing Occupancy Rates in the United States by Type of Structure and Age of Household Head, 1960-1970

Year	Housing Type	Age of Household Head							
		15-19	20-24	25-29	30-34	35-44	45-54	55-64	65 +
1960	One-unit structure	47	55	68	75	79	78	75	73
	Multi-unit structure	47	41	30	23	20	21	24	26
	Mobile units	6	4	2	2	1	1	1	1
	Total	100	100	100	100	100	100	100	100
1970	One-unit structure		33[a]	57	72	79	77	74	67
	Multi-unit structure		59	38	25	19	21	23	30
	Mobile units		8	5	3	2	2	3	3
	Total		100	100	100	100	100	100	100

SOURCE: William Alonso, "The Population Factor and Urban Structure," Joint Center for Urban Studies, 1977, adapted from Thomas C. Marcin, "The Effects of Declining Population Growth on the Demand for Housing," U.S. Department of Agriculture, General Technical Report NC-11, 1974, 5.

a. This estimate is for all households under 25 years.

housing stock available to the working and welfare poor will now be sharply diminished, squeezed between reduced rates of filtering at one end and the childless multiworker household at the other. There is, in this, an incipient class conflict between the new, young, well-educated professional class, actively pursuing alternative living arrangements and lifestyles, and the majority of the children of working-class Americans, for whom marriage and the home in the suburbs remain a desirable goal: Harris polls continue to report that of the 35 percent of American dwellers who plan to move in a 2 to 3 year period, 53 percent of them plan to move to a suburb or to a rural area. Indeed, one may argue that the main value struggle in the United States today is between liberal-regarding upper-middle-class intelligentsia (a minority of whom are our most vocal Marxists) and a newly middle class and more conservative proletariat, which rejects both "alternative" lifestyles and the left wing's egalitarian arguments. For the intelligentsia, material goals have been succeeded by those of the quality of life and of self-actualization (for the Marxists, notoriety and in-group admiration); for the middle-class workers, however, material welfare and economic progress remain the dominant concerns.

Is it possible, then, for private-market revitalization to be sustained? A variety of factors suggest otherwise. First, the baby-boom generation will age, and is followed by much smaller age cohorts (Table 7.7). The population bulge, as it ages, will continue to disrupt one national institution after another. In the 1950s and 1960s it created problems of expansion for the public schools and universities—institutions which more recently have had to cope with the ordeal of shrinkage as their user populations have subsided. When the population crest reached the 18-24 bracket, it multiplied crime rates and redirected national job-creation efforts to the alleviation of youth unemployment. Perhaps the greatest adjustments for public policy of all kinds lie ahead, when the babies of 1950 become the aged of the year 2015. Among those adjustments, as before, will be those in housing preferences, causing the demand for inner-city living in revitalized neigh-

TABLE 7.7   Five-Year Changes in the U.S. Population by Age Class, for
Population 20 Years and Older, 1970-1990 (in thousands)

	Age Class					
Period	20-25	25-35	35-45	45-55	55-65	65 +
1970-1975	2,208	5,902	- 375	334	1,131	1,433
1975-1980	1,640	5,720	2,706	-1,153	1,284	2,114
1980-1985	- 751	3,819	5,807	- 319	328	1,814
1985-1990	-2,514	885	5,624	2,638	- 990	1,993

SOURCE: Alonso, op. cit.; computed from U.S. Bureau of the Census *Statistical Abstract*
projections.

borhoods to subside rather than increase, at least until the early
twenty-first century. Additions to the supporting job base may
not be there either. There appears little likelihood that new
post-industrial employment clusters will develop in highrise
office complexes in inner-city locations in the near future as
they did in the 1960s; headquarters decentralization now appears the stronger force. Further, as the cohort ages, mobility
will decrease. For all of these reasons, the prospect for wider
inner-city revitalization appears to be bleak, unless continuing
recession cuts deeply into the new housing industry, and if that
comes to pass there will still be an inner-city problem of the
displaced and the disadvantaged.

There are other reasons, too, that we might expect the
dilemma of the inner city to continue, and even to deepen. The
regional income convergence discussed earlier was a product of
the extraordinary factor mobility that enabled new growth
opportunities to be exploited and growing regions, in turn, to
become increasingly autonomous via import substitution. The
branch plant movement, for example, has been one of the most
potent sources of manufacturing job decentralization from the
Northeast since 1950. Declining mobility suggests the
likelihood of progressive regional income divergence in the
future. Already the signs are there, for nominal income convergence, in fact, does imply real income divergence to the
disfavor of the major cities in the snowbelt heartland. Costs of

living and the burden of negative externalities appear to be higher in the inner cities of the older urban-industrial regions, whereas real or perceived environmental amenities are greater in the sunbelt and in the intermetropolitan peripheries. Lessening factor mobility would imply a decrease in autonomy and a commensurate increase in commodity trade. Recall that the Heckscher-Ohlin theorem postulates that when factor endowments and prices vary, the response is regional specialization and commodity trade. The Siebert corollary states that if factor prices are equalized via factor mobility, the result is regional income convergence and autonomy via import substitution. Myrdal's dilemma was one of the deepening factor endowment differentials and of progressively more profound heartland dominance of poverty-stricken hinterland economies. Mobility produced the equalization of Siebert's corollary. Increasing immobility can only, then, lead to hinterland dominance of increasingly poverty-stricken heartland economies. Indeed, the balance of political power in the Congress has already shifted in this direction.

There is no reason to believe that aging industrial cities will be able to revitalize unless they are able to develop a postindustrial high-technology or service activity base. Neither is there any reason to believe that those metropolitan regions developing such a base will do so in a manner that re-creates the inner cities of the past. Unless there is a prolonged recession, there is little basis for believing that private-market revitalization of inner-city neighborhoods will diffuse much further. There is every reason to believe that settlement patterns that have emerged in the past decade will continue to diffuse and differentiate.

The continuing public policy problem will then be that of ameliorating the heavy social costs incurred by the concentrations of captive individuals without access to the real economy, concentrations that continue to be characterized by high unemployment rates, especially among minorities, and by low educational achievement, drug addiction, crime, and a sense of hopelessness and alienation from society. Attention to causes rather than symptoms demands that factors that exclude individuals from the mainstream of society and from meaningful work opportunities be of prime concern, for work is a measure of worth in the United States. Such factors include poor skills, cultural gaps, language barriers, and race, and may have to be addressed by both law and remediation. Programs aimed at creating work situations, intensifying quality educational assistance, and improving health and nutrition so that children will be able to learn, are essential. Many such efforts at human capital development must be people-directed, but a deliberate strategy of fostering geographic as well as social mobility should not be excluded.

A restructuring of incentives played an important role in the increase in homeownership and the attendant transformation of urban form after World War II. There is no reason to believe that another restructuring could not be designed to lead in other directions, for in a highly mobile market system nothing is as effective in producing change as a shift in relative prices. There is, then, a way. Whether there is a will is another matter, for under conditions of democratic pluralism, interest-group politics prevail, and the normal state of such politics is business as usual. The bold changes that followed the Great Depression and World War II were responses to major crises, for it is only in a crisis atmosphere that enlightened leadership can prevail over the normal business of politics in which there is an unerring aim for the lowest common denominator. Nothing less than an equivalent crisis will, I suggest, enable the necessary substantial inner-city revitalization to take place. Until that crisis occurs, dispersion and differentiation will prevail. Some limited private-market revitalization will continue, to be sure, but within a widening environment of disinvestment manifested geographically in the abandonment of the housing stock put into place by earlier building cycles.

8

FUTURE ORGANIZATION OF THE EUROPEAN INDUSTRIAL CITY

"Urban Alternatives"

STEPHAN JONAS

This essay deals with a research field that depends on several hypotheses. These may be summed up in three points:

- The sudden decrease of the economic growth of the richest capitalist and socialist countries, from 1974 onwards, has just as suddenly questioned the so-called American urban growth of the cities and larger metropolitan areas of Europe.
- The final stage of this wild, expansive, and almost unregulated territorial urban growth, which has generated anticity movements (the suburbs and the urban sprawl) and tendencies to urban dissolution, has become of great concern to decision makers, urbanists, and sociologists. It leads to a drastic reexamination of the dominant urban policy, raises plans for original solutions—which are minority though novel, and partial though experimental—and for "substitute urbanism" doctrines *(urbanisme de rechange)*.[1]
- Similar to the situation in 1900 and the beginning of modern urbanism, this period of economic and social and urban crisis leads to an intermingling of urbanistic and political programs. From this point of view, it is interesting to analyze the politization[2] of the substitute urbanism plans of two main

political trends: the neo-*liberal* trend in France and the *socialist* and *communist* trend in Italy.

This crisis in city growth and urbanization in Europe is not new,[3] and could, up to a certain point, be compared to the crisis of the European "metropolis" at the dawn of this century, when some architects, philosophers, sociologists, and social investigators, such as Raymond Unwin, Ebenezer Howard, E. Henard, J. Stuebben, G. Simmel, J. Sombart, M. Weber, and others who were interested in urbanization phenomena observed the outcomes of the wild industrial growth in the larger European or American urban areas.[4] Have the main themes of a substitute urbanism changed since 1900? Very little, indeed. Today, as at the beginning of this century, they encompass the significance and meaning of the city, of the urban life, of the various units of social life (neighborhood units, districts) that develop or should exist within them, the issues of historical centers, the very essence of which is jeopardized, and of rural-urban and work-residence relationships.

The European industrial city is shattered. Ever since 1945, and even as recently as just a few years ago, our decision makers and experts showed a glorious image of the rapidly growing European cities that, according to them, were to double the number of their inhabitants every ten or fifteen years. Some notable sentences of the great Italian Renaissance thinker have been forgotten: "Nobody should believe that a city can expand indefinitely.... As for me, I shall say that the growth of cities originates partly in generative abilities of men and partly in nourishing abilities of cities."[5]

The European urban growth as a whole has come to a critical point, where functional specialization, giantism, and obsolescence more and more remove any orderly image of urban unity and totality. Through the disintegration of the industrial city—either into a huge polycentrical metropolis or, on the contrary, into its scattering in disurbanized areas—the very nature of urban pattern and urban life is questioned.

What about the historical urban heritage? What about the nineteenth-century industrial city in this fantastic European network of the "thousand cities," an archipelago that is unique in the world? Essentially made up of historical centers that are still holding out, though they are artificially deprived of their dynamism, their autonomy, their specific and original features, is this network "worn out"? Is it to disappear, together with the industrial cities? Is it to give way to new metropolitan systems imported from America or Japan?

These questions also directly relate to the future urban organization of European industrial cities, insofar as the slackening or the halt of urban growth may generate a stagnation policy for cities, without a change in the traditional type of growth (wild and expansive), as well as a limited expansion of cities that thwarts the development of the nineteenth- and twentieth-century urban web. But this conception of a slackening growth may also result in the appearance and assertion of "extreme propositions," minority yet creative propositions—at least as far as ideas and partial experimentations are concerned.

The Council of Europe, one of the most active centers of reflection on and studies of urbanization in the Old World, has for over fifteen years led a cultural action promoting the protection and conservation of the European urban and architectural heritage. Recent stands taken in this action show that the member states are determined to face the urban challenge. As a matter of fact, they have for the first time decided to organize their action within the frame of a new European policy, one of territorial planning, of national development (*aménagement du territoire*).

In 1978 in Vienna, a European Conference held on a ministerial level announced that the first draft of a European charter of national development would be made in time for the next conference of the Ministers in charge of national development. To promote this charter, a European campaign on urban renewal would be launched by the Council of Europe.[6]

But already a few years ago, and probably under the influence of the crisis that started in 1974, several symposia and confrontations that were organized by the Council of Europe— mainly since the 1976 Confrontation of Berlin on the larger European cities[7]—rightly stood against the idea that the European network of historical industrial cities should be doomed to disappear. On the contrary, in October 1978, in Ferrara—the hometown of Biago Rossetti, the first urbanist of modern Europe, and the high place of urban culture in the Old World—urbanists and architects of European countries considered this historical network as the only chance for an urban rejuvenation, for human-sized cities and urban life, in Western Europe.[8]

Starting from these general points, I shall submit two doctrines of urbanism *in statu nascendi* that are relatively new and creative and seem to me quite significant of the tentative efforts following the slackening of urban growth in Europe:[9]

- *A doctrine of "soft" urbanization,* ecologically oriented, that advocates an environmental policy for the five years to come: It is a French-originated doctrine, within the neoliberal circles of the central power that preceded the Mitterrand government.
- *A doctrine of "alternative" urbanization,*[10] of a social nature, that advocates a policy of a social urbanism for the next few years and in the short run. This doctrine is of Italian origin, arisen in socialist-communist spheres opposing the central power, on a municipal and a regional basis.

TOWARD A "SOFT" URBANIZATION POLICY?

In the near future, plans for protecting the European cities and their historical and modern heritage—cultural, social, and local—will undoubtedly have to face the neoliberal theses of urbanism, the vitality of which has astonished many European sociologists. These neoliberal theses of urbanism are based on an approach of the ecological type on the ideological level, and on a prospective and futurology approach on the methodological level. In the field of national development, a new, all but official, French doctrine was being set up; it was based on an environmental policy, on a "soft urbanization" (a phrase that was bound to make a hit in Europe). Furthermore, it seems that the former president of the French Republic, Valéry Giscard d'Estaing, was personally involved in the elaboration of this new doctrine.[11]

The origin of this doctrine of urbanism can be traced back to the French works of various prospective research groups, which arose on one hand within organizations of development, such as the Délégation de l'aménagement du territoire et de l'action régionale (DATAR) and the Commissariat général au plan, and, on the other hand, within the various departments in charge.[12] During 1975-1976, several important reports were issued, and the French government widely spread them. These reports broadly benefited from the works and methods elaborated by the French tradition of prospective research work. I shall only quote the most important ones in my opinion: the Nora-Eveno Report on old housing, the Lion Report on public housing, the Delmon Report on associative life and participation, and the La Brousse Report on time planning.[13] The contents of some of these reports and also their prospective approach even succeeded in influencing the policy and the moves of the Council of Europe and of the Common Market.[14]

As far as the economic and social policy is concerned, this new (was it really new?) doctrine of environment refused actually to question the economic growth from 1945 on. But it readily admitted that this growth needed a mutation, for it was too fast. Consequently, the liberal economy, which was moreover to be concilated with the protection of the natural and cultural heritage, was to yield to a softer-growing economy.

Furthermore, this neoliberal policy was meant to be very European and international with regard to environmental matters. It readily admitted that the accelerated integration of the economies of the member states led them toward an environmental policy on the scale of capitalist Europe. The main directions for reflection and action of this new policy were quite accurately recorded in the Charter of Quality Of Life, and they may be summed up in five points:

(1) town planning
(2) protection of the natural heritage
(3) fight against pollution, nuisance, and waste
(4) time planning
(5) reinforcement of the part of associations and guarantees of the right to information

As far as town planning is concerned, the new ecology-oriented policy of urbanism, which claimed to be of human size, would mainly be organized in the following directions: local democracy, open-space land policy, traffic, urban aesthetics, housing policy, and public facilities. The lack of balance the twentieth-century city bears in many ways should progressively be rectified by the seeking for quality of life, in order to turn the city again into a place for exchange and meeting. The urban aesthetics direction reminds us of the conservative traditionalism that imprints this trend.

Through protecting the natural heritage, the point would be to introduce into the French plans a policy of qualitative national development that would turn toward a policy of protecting nature. These concerns for active protection of the territory were not actually new. What was new though, was the integration of this policy into a wider, more ambitious ecological project.[15]

In 1975 the program for fighting pollution, nuisance, and waste was one of the four directions of the neoliberal policy that superseded the Gaullist power, in the field of quality of life. The other three directions were the development of associative life, time planning, and reform of the public benefit investigation (*enquête d'utilité publique*). Ever since, this program has included the struggle against waste, and a policy for energy saving—a sign of this time and of the adaptive ability of the neoliberal theses.

The time planning policy may be the most original French suggestion, and the works on large-scale implementation had actually already started in the early 1970s.[16] Starting from earlier American and Russian works on time budgets that had been looked up in Europe, as early as 1964, by the European Center of Coordination of Social Science Research (Centre européen de coordination de la recherche en sciences sociales) in Vienna (a UN affiliated organization), the French prospective research groups produced their first report on time planning in 1972, for the French government (the Chalendar Report).

The axis and ideology of "spare-time civilization" has essentially shifted to the symptomatic choice of "labor-time planning" in the present neoliberal proposals—which constitutes a new policy of time planning in France (see J. Du Mazedier on his "spare-time civ." concept).

France has now been given an overall scheme of time planning, covering problems of conventions for labor-time flexibility, for changeable schedules as well as everyday urban leisure, and for planning of late opening hours for socio-cultural facilities. The implementation of the principles of time planning in our cities would obviously bring forth considerable changes within the urban space.

As for the policy toward associative life and participation, it should be considered as one of the best points of the neoliberal doctrine. The neoliberal social regime wants above all to be a "society of communication and participation," in which the people-community types would be more favored than the mass-organization types. Therefore it is out of the question for the neoliberal doctrine to consider participation and associations as new social forms of urban, rural, and local representation, as in Italy. Rather they are to be thought of as a *relay,* as a questioning partner for central power to decentralize decision.

The governmental report called the Delmon Report is entitled "La Participation des Français à l'amélioration de leur cadre de vie" and was launched in February 1976. It has not met a fortunate fate: Poorly defended and supported by the minister in charge, the 45 associative proposals were drastically cut down by the Parliament, which finally maintained only twelve of them, among the least important ones.

*   *   *

The main statements of the neoliberal trend meet those of several European trends that attempt to analyze the present crisis; but their process is specific and their choice sets down priorities that are determined by general considerations on environment and ecology. This trend ascertains that:

- the French and European cities have grown too quickly;
- the rural exodus has been too fast;
- growth has resulted in too much pollution, waste, and nuisance;

- the historical centers and the urban network have been invaded by American-style giantism and anonymity;[17]
- economy has been neglectful of ecology; and
- ecological inequalities do exist.

However, the short-run aims (five years) were not selected according to the general data of analysis, but according to the general neoliberal outlook that the resulting doctrine outlines as the urban society to come:

- substituting quantitative accumulation by a softer growth that is more qualitative and more social;
- fighting giantism in town planning;
- setting up a policy of protected spaces in order to preserve (rural) nature and (urban) environment;
- rejecting waste, time splitting, and anonymity;
- advocating dwellers' participation for the improvement of everyday life;
- reinforcing the part of associations for a better way of life (*associations de cadre de vie*);
- expressing concern about quality of life and wider equality for it;
- reducing ecological inequalities and forwarding a policy of an ecology that is independent of any ideology; and
- recognizing the need for a new and strong impulse for applying the European environmental policy.

Like all urbanization doctrines and theories that do not regard space as politically neutral, the neoliberal theses on soft urbanization refer to the quite specific aim of the "advanced" liberal policy drawn up long ago by the Giscardian trend. The opposition, including Gaullists, cast suspicion on the authenticity and originality of these theses by charging the liberal trend with "giving itself credit" either for the ecological theses set up by the independent ecological left, or for the Gaullist idea of participation. Even if this were partly true, I would stress the fact that in this trend a relatively coherent doctrine has been elaborated, which tries to get enforced. Therefore, I should find it wrong to consider this environment-and-participation doctrine only as a mere political by-product in the political sense of the word. And I think this trend of thought, this postwar liberalism, will have specific urbanistic features and will influence the French and European national development policy even in the short run.

TOWARD AN "ALTERNATIVE" URBANIZATION POLICY?

Among the urbanization theses based on the necessity of forwarding a new model of urban development, instead of the traditional urban development based on continuous territorial growth, the new left Italian doctrine, called *urbanistica alternativa* by some of its conceivers, deserves special attention. In its conceptual form, this doctrine in the making came into being as early as 1975,[18] and its opponents already consider it obsolete; as opposed to the French neoliberal theses, it was set up in a radical opposition to the boundless, uncontrolled expansionist traditional urban growth, as it appears particularly in the metropolitan pattern of growth. The Italian "alternative" urbanization is characterized by an approach of social urbanism on the ideological level, and by a planning approach on the methodological level.

The Italian doctrine of socialist-communist urbanism was born during the 1970s and has a threefold origin:

(a) the failure of the Italian cities to solve their problems of growth;
(b) the Italian institutional change resulting from the reform of regional decentralization; and
(c) the powerful urban social movements that were created in the past ten years.

This urbanism doctrine in the making is also very European and international in the sense that its unfolding into a spread-out and experimental theory probably owes a lot to the cultural action of the Council of Europe toward the defense and protection of urban and architectural heritage.

About the defense of European historical cities, this doctrine has raised a first interest in Europe, although a qualified one, because of its radical and sometimes Marxist stand. It follows the experience of Bologna, which has been a "model city" for the Council of Europe since 1974, as an implementation of the cultural policy of defense of the European urban historical heritage.[19] The Bologna experience has been an important stage for this trend insofar as, during the Second Confrontation of the Council of Europe, which took place in Bologna, and through the very experience of the city, the necessity was discussed of regarding the defense of historical centers in terms of urban unity and general development of the urban area and its territory. On the other hand, the city of Bologna decided to put an end to the demographic growth of the city, thus directly questioning the expansionist policies of the central power and of other European cities.[20]

The idea of an alternative urbanism as a criticism of territorial growth has already been expressed and drawn up specially by the Italian professor and architect Leonardo Benevolo, who, at the Confrontation of Berlin said:

> Today, the historical center is restructured from the periphery, tomorrow the periphery will perhaps be restructured from the historical center; in fact the aim is not the development of a specific and privileged district within the city, but a way of conceiving the totality of the future city, so that it may really be said "modern." ...The unremitting destruction of the inner core and the continuous growth of the periphery are the two interdependent components.... The scheme of preserving the historical center is part of a development plan, that is an *alternative* to the previous one and completes the *limitation of peripheral growths.*[21]

These experiences and ideas of "alternative" urbanization stirred up an acute interest in France, both within the circles of the left opposition and in the highest places of the political power of the government and the state. The cultural actions of the Council of Europe provided opportunities. But the neoliberal circles went further: They had several Italian papers translated, and the leaders of the "model cities" were invited to come to France to talk about their experiences. The instance of the Mayor of Pavia, Elio Veltri, and his influence, and more generally the development experience of Pavia, are very symptomatic from that point of view, and show some aspects of the policy of adaptation of the French neoliberal trends.[22]

The alternative urbanization theses rouse a sharp interest in several political circles of the Council of Europe. Thus, the main aforementioned proceedings of the Confrontation of Ferrara, the impact of which I have already stressed, are for almost all of them documents that convey quite well the theoretical and developmental problems that this trend sets.[23] These theses are also beginning to raise interest in some urbanism and architecture teaching circles.[24]

Where the economic and social policy is concerned, this doctrine in the making radically questions the national development state policy in Italy, as in the remaining capitalist and socialist European countries still based on the continuous growth of cities and urban networks, on a hierarchical—not an interdependent—pattern. This is expressed and symbolized in the metropolitan areas of North America. Since 1945 this system of urban growth has brought influences deemed disastrous, especially in the following directions:

- destruction of the balanced network of the older urban frame, forwarding the hierarchical urban system of the metropolises;
- discarding of the hinterland and the agricultural crisis in relation to the city;
- birth of a pattern or urban development that has favored the quantitative growth of the cities, instead of quality of life and "human size";
- disruption of the unity of the city, with the historical center being isolated from the city and the periphery underrated compared to the city center;
- isolation of the city from its territory;
- lack of balance in the structures of the historical center: sharp increases in white-collar occupations, disappearance of production, and decrease of lower-class housing; and
- private reapportionment of urban land: land speculation and decrease in social facilities (*équipments collectifs à caractère social*).

According to this trend of thought and action, only one efficient remedy is supposed to exist against the continuous urban growth in its territorial expansionist form and the connected socioeconomic phenomena that prejudice lower-

class people and urban life; the radical and structural transformation of the whole urban development mechanism. The
strategy of radical transformation of the traditional urban
development into an alternative urbanization of the city and its
territory would not do without a protection policy. Campos
Venuti suggests to sum up this protection policy within five
points:[25]

(1) public protection
(2) social protection
(3) protection of production
(4) protection of environment
(5) protection of planning (programmation)

Except for slight differences which I shall not analyze in
detail, these five types of protection accurately express the main
concerns of this trend, and would summarize the "overall
project of the urbanist alternative for cities." Moreover, they
would allow for an active protection of the essential factors of
urban life. The main target of the Italian socialist-communist
pattern of the alternative urbanization type is to reorganize the
city, not to increase it, a priori, by territorial expansion,
whether it be slackened, as intended by the French new
neoliberal growth policy, or of the traditional expansionist
type. In fact, the Italian alternative pattern favors a qualitative
urban growth, which turns toward the inner-city space, toward
the heart of its historical core—including the nineteenth-
century development—instead of turning toward the suburban
areas, the outskirts of its territory, its hinterland.

As for the public protection, it would essentially encompass
community use of urban lands, to fight and check the private
appropriation process of the Italian cities, a process that had
been increasing considerably during the expansionist postwar
period. On the other hand, the collective tendencies of urban
life were to be reinforced: various phenomena of collective
reappropriation, such as participation, associative life,
neighborhood committees or unions, give an opportunity for
reappropriation of urban spaces by the users and the dwellers.

The important sociological concept of reappropriation should well be taken notice of in this context. This concept was already drawn up, or at least discussed, during the 1974 Confrontation of Bologna of the Council of Europe. The protection of public goods also commends the increase of collective sociocultural facilities (*équipements collectifs socio-culturels*) that promote gatherings and social life. That would mean setting up a new policy and a new conception of services and public facilities that would have a less rational but a more communitarian purpose.

The social protection would essentially aim at maintaining within the urban and historical centers the traditionally urban native lower classes, workers, craftsmen, shopkeepers, retired persons, whom the expansionist urban growth and land speculation tended to drive out and expel toward the peripheral areas of the cities. The point would be to fight any tendency to residential segregation by strata or classes by appropriate means of a social type of urbanism.

The protection of production would reverse the present tendency to drive away and expel from the central areas of cities the places of production, even though they are compatible with a central settlement—nonpolluting light industries and handicrafts—in order to replace them by white-collar activities, which are more and more obtruding and speculative. The principle of protection of the production in an urban area also concerns—and this is very important—the agricultural areas that are located in the immediate surroundings of the city and must be preserved from the excess of planned urbanization, as well as from land and bank speculation, originating in a recreational capitalist policy. With the defense of the bordering agriculture by a regionalist conception of the rural-urban relationship, the Italian trend of thought directly follows the socialist and Marxist historical tradition, which considers the rural-urban relationship in egalitarian and dialectical terms, and not as a hierarchical relationship favoring the city.

The protection of the environment, both natural and architectural, means a new policy to be drawn up, for an active defense of historical sites, built-up, and open spaces, on one hand, and for the conservation of public housing in the central areas, on the other hand. But to protect the natural environment is also to democratize the right to natural spaces.

The protection of planning raises the basic question of the need for elaborating democratic urbanism plans—as opposed to the technocratic ones—while fundamentally reexamining the present conception of urban planning under the state authority. Plans produced during the period of boundless growth have been understood as the technical implementation of understructures, not as a means for a consistent urban development (space, regulation, transportation, social relations), which proceeds according to priorities.

This critical questioning about the part of urban master plans, and, more generally, the part of a democratic planning for the city development, is essential, and relates to two theoretical problems: first, a too hasty belief in the efficiency of a scientific and technological revolution, without sufficient awareness of its relationship to resources, and its possible occurrence only through a thorough social change; second, an excessive trust in the abilities of experts and the established authorities, without the users, whether individuals or groups, being given a chance to participate in conception, decision, and implementation.

*   *   *

The project of alternative urbanism is also a methodological project. It sprung from the esprit du plan, and therefore it is close to prognostic, social prognostic in particular, after the Marxist inspiration and tradition. The case of the urban development of Ferrara is an example of a complex planning project of the communal territory, where a triennal scheme of economical and social development already exists and was conceived in a close connection to the urbanistic scheme (Commune di Ferrara, Il Piano triennale degli investimenti 1979-1981, Ferrara 1979). But this project implies doing field work and taking into account the progressive urbanistic experiences on the spot, rather than conceiving programs of action after the prospective method, such as it is used nowadays in many futurological studies approaching the future through speculation and sheer theory. The conceivers of that doctrine in the making like to stress this specific feature of alternative urbanization: It has derived from a slow accumulation—of at least fifteen years—of a local democratic institutional urban experience, of a large participation between municipalities, technicians, and consulted and involved inhabitants. In this respect, alternative urbanization is also a new associative form.[26]

Thus, the primary thesis of this trend amounts to this: In order to avoid the traditional expansionist growth, many Italian cities took significant urbanistic steps, limited but efficient steps that on the whole provided the experimental and theoretical framework of the urbanist alternative. That is important, and reminds me of a fantastic sentence of Marx when he spoke about the social importance of human praxis and post festum experiences about knowledge: "They do not know, but they do it." Which were the steps taken? They can be summed up in the following points:

- limiting excessive growth prospects and reducing residential high densities;
- allotting the best urban spaces with adequate inner cores to lower-class housing construction;
- social and architectural defense of historical centers;
- measures against "pathological white-collar job increase" in the city centers;
- steps against the speculative expulsion and the move of industries and handicrafts from the central urban web toward the peripheral areas;
- drawing up of criteria for the recovery of obsolete housing;
- reservation of interurban territories for future sociocultural facilities;
- keeping the native inhabitants in the historical centers;
- through appropriate restoration conventions, involving the landowners' participation in urbanization costs;
- bringing the city back to its "human size" by restoring its potentialities as a center for gathering and social solidarity;

- fighting the progressive and continuous decay of everyday living and working conditions;
- controlling urban policy by a new form of people's participation in public life;
- modernizing overweighted and disrupted supply networks;
- limiting highway traffic prospects and organizing public transportation; and
- preserving natural space and agricultural land around the city.

On the methodological level, this trend of thought and action finally represents an older European tradition of urbanistic practice, dating back to the nineteenth century: considering and drawing up proposals of social prognostic for urbanism—in the sense in which it has been understood by the socialist trends(not by the social democrats) for about a century— starting from socially significant local experiences. The urbanistic experiences of several Italian model cities led by a left coalition, such as Bologna, Venice, Pavia, Ferrara, Livorno, Modena, or others, are of great value, as they do not accept the belief that historical industrial cities are done with and must dissolve in a territorial unity of some sort, or dislocate into something uncontrolled and diffuse.

Of course, all problems of the survival of our European urban network are not solved by the Italian trend of an alternative urbanization. It is true that the experiences of a city confederation in the regions of Lombardy, Venetia, and Emily-Romagna have only just begun. Some normative aspects of planning—mass mobilization, strategy against landowners and estate agents, municipal interventionism, neighborhood committees often more or less tamed by municipalities, and so on—do not always seem convincing or democratic. But it would be wrong to see in this new doctrine in the making a left municipal political endeavor to get hold of the cities only to administer them. With these theses of alternative urbanism, we are confronted with a modern doctrine of urbanism, a genuine supporter of the historical social and architectual, ecological and cultural heritage of the European cities that are going through a critical time.

SUBSTITUTE URBANISM DOCTRINES:
CONTINUITY AND RENEWAL

The main aspects of the up-to-dateness and significance of the substitute urbanism doctrines that I have submitted here may be arranged in four themes of reflection.

(1) urbanization and history
(2) urbanization and politics
(3) urbanization and public opinion
(4) urbanization and science

The relationship between history and the present resurgence of substitute urbanism doctrines seems to me significant in two respects. First, the origin of the main political, scientific, and intellectual forces that constitute the basis of the recent actions toward a new urbanism, lies in prohistoricist trends, concerned about the past and the historical cultural identity of cities. From this point of view, we are in a significantly different situation from that of the 1900-1930 period: In that exceptionally rich period of Europe, the urbanistic experimentations and critical reflections about the European Grossstadt were largely influenced by the vanguard antihistoricist trends—whether expressionist, functionalist, futurist, or constructivist—observing a sort of religion of the present. On the whole, except for the theories after Fabian socialism (Unwin) or Fourierism (T. Garnier) in 1900, these trends set up theories and doctrines of urbanism tinged with universalist conceptions that are too simplistic and usually ignore the diversity of local, regional, and even national urbanistic problems. Can the present prohistoricist trends reduce their influence?

Now, is the present concern about historical continuity deep and real? The neoliberal and socialist-communist trends I have studied, submit to the present campaign of protection of the historical heritage of cities more than they invent or inspire it. The influence of time-budget studies on the neoliberal trend and of planning on the socialist and communist trends shows that within these trends the present research on new urbanization rather relies on methods and conceptions inspired from the future: futurology and prognostication. These trends undergo the influence of that kind of religion of the past that gradually prevails instead of the postwar pseudo-modernism, which was so damaging and destructive in the European industrial cities.[27] Their visible signs are well known: mummified historical neighborhoods (e.g., "listed") or, on the contrary, sites that are renewed and emptied of their lower-class strata; fake pedestrian and comparative shopping streets; historical prestige buildings restored in situ in the midst of obsolescent areas; and so on.

The great novelty of the present rising of substitute urbanism theses lies in the repolitization of the urban issue. I agree with Professor Benevolo and with Professor Freund that we are still suffering from the lack of studies on the relationship between politics and urbanism.[28] Since the 1920s and 1930s, and because of the ebbing of the revolutionary period followed by the general rise of Fascism and Stalinism in Europe, modern urbanism and the urbanization doctrines and theories have been isolated from the political democratic debate, and since the 1960s, they have become a matter of ideas, political opinion, praxis, and committed experimentations—but no longer a business for technicians, a matter of technostructure, and sheer technique theory, exclusively dependent on the established power, whether liberal, conservative, or socialist.

The time of repolitization of the urban issue bears an original feature, compared to the short but rich periods around 1848 and toward the end of the nineteenth century, dominated by the Young England and William Morris, and also Raymond Unwin and the Fabian socialism. This feature probably lies in this fact: To the discovery of the essential necessity of confrontation, that all political regimes must carry out between space planning and socioeconomic planning, is now added the necessity of confrontation and negotiation between the socioeconomic policy and the environment and resource policy—raw materials, agriculture, energy, open space land, and architectural and natural environment. The Italian theses of alternative urbanism are significant and important in this respect.

Regarding both political and historical grounds, we may be witnessing a reversal of tendencies between the parts respectively played by the liberal, conservative trends, and the socialists and communists, in matters of urbanization policy and theory during the nineteenth century. In the philanthropy inspired by Catholic and social Protestant liberal trends, liberalism had its own theorists, social movements, and trends of thought that conceived and carried out urban and housing experiments, which were certainly not overall but were real and important. The liberal trend has lost this historical role of elaborating reform plans. The socialist and communist trends are now to play that part fully, at least so far as management of municipal power is concerned, in the framework of an overall policy of pacific coexistence, where the existence of capitalism is acknowledged. The Italian instance is quite significant here. But the French instance might show that neoliberalism wants to claim and play a similar part, though relying on the established power and state planning.

I presume we all agree on the dominant part that the movement of democratization, the birth of public opinion, and more generally the rising awareness of the need for a democratic community have played during the nineteenth century, in favor of the birth of modern urbanism. Do we not find the same phenomena in the present questions on the cities in crisis and their chances of survival? Are they not influencing our doctrines and theories in a very active and creative way? Here I refer to two particular forms of these phenomena, which have emerged again in Europe since the 1960s, after thirty years' absence: participative ideas and practices, associative movements and phenomena that are essential parts of both doctrines I have analyzed. The French neoliberal doctrine I have submitted, which is planning a reform of the 1901 Act, the basic act for associations in France, is mainly oriented toward the phenomena of "overgroups" (*sur-groups*) as associative movements on a national scale, and among them towards "ecologists"—or rather toward their ideas—because of their political impact and their suggestions on environment. The remaining associative movement on the local level is not ignored, but only regarded as a "relay," a possible municipal support, and not as a new potential social form of local urban representation. This is the response of politicians to a trend stemming from the centralized state power, which considers that the state administration is the main warrant for the preservation of common urban and environmental interests.

The Italian socialist-communist doctrine is an interesting model which has succeeded (at least to some extent) in integrating participation and the associative movement into its program. It actually is a new potential social form of local urban representation. After integrating neighborhood committees as propositional (yet not decisional) organizations into municipal management, this model gives heed to the cooperativement movements.[29] These are self-managing and associative forms referring with relative loyalty—even today— to the Rochdale principles,[30] movements that are reviving in Europe, in the countries led or influenced by social democracy, after the acme of the 1920s. In the basic urban and housing questions, this is undoubtedly a positive evolution that may be of great consequence.

I lay stress on the associative phenomena, as the expression of an ambition for a democratic community, in spite of the ambiguity of associative movements in Europe and the ideological influence of middle classes and intellectuals on the present rising of this movement.[31] With the influence of associative ideas and practices, the substitute urbanism doctrines get a share in urban utopia: an attempt to replace (Italian pattern) or to conciliate (French pattern) competitive capitalism with an association-rooted capitalism. Here, the concept of association is meant in the Saint-Simonian sense.

In Europe, the quarter of a century that followed World War II brought with it some flourishing scientific urbanistic works based on American-like empirical and quantitative methods and theories, inspired by the technical and scientific revolution, and also by a genuinely European formal aesthetism. One might believe that it could have a negative impact on the role of science in the field of urbanism. This is not the case, at least as regards both doctrines under discussion. The thorough scientific studies of the Italian model are strikingly momentous;

advocating for the past, this trend is definitely oriented toward the future and trusts in human practice, in the Marxian and Lukacsian sense of the phrase (*menschliche Erfahrung*).[32]

In the French neoliberal model, the part of science and, a fortiori, of science of the future is widely acknowledged, but the idea of a necessary control of science is a characteristic feature of this trend.[33] Besides, its conception of the future proceeds from the future, rather than from the present, and the following sentence of the French thinker Gaston Berger, the father of prospective thought in France, could be its motto: "Prospective is neither a doctrine nor a system. It is a reflection on future that endeavours to describe its more general structures and tries to define the elements of a method, which could be applied to our accelerating world."[34] In this conception, we can obviously detect the influence and the prospects of the larger French and international scientific associations, of the European bourgeoisie for futurology and prospective, such as the Club de Rome, the French sections of the Hudson Institute, the Congrès européen de la culture or the Association internationale FUTURIBLES.[35]

*   *   *

In this essay, I have raised questions about the present interest and the rising of substitute urbanism doctrines and theories in Western Europe, in a crucial period in which, on the European level, we can detect a sharp awareness, an urge to stop the traditional urban growth, the anarchical and quantitative territorial expansionism of our cities.

This question is actually essential in Europe, where the postwar growth crisis culminating in 1974-1977 has brought the traditional urban policy of a diffuse and continuous growth in the wealthier capitalist European countries to a sudden stop. But once that stage has been reached, the leaders of these countries, whether social democrat, liberal, or conservative, do not know exactly what to do and what to undertake afterwards: to grapple with the structures or to wait for the trouble to die out, and start all over again as before.

We live in a time—and it is difficult to say how long it will last—that is propitious to the genesis and generalization of substitute urbanism doctrines and theses which go as far as suggesting a structural change, without actually questioning the foundations of capitalism and the dominant power.

By selecting two doctrines of substitute urbanism in the making connected to important European political groups that are ideologically quite different, and furthermore relative to a different national reality, I did not mean to carry out a comparative study nor a typically intellectual reflection that is fashionable in our academic spheres. Such would consist in looking for formal and conjunctural common features, thus implying that they are reconcilable, that they may be drawn together and connected—an idea many politicians have been dreaming of in France, Italy, and elsewhere.

For a sociologist like me, it is clear that the liberal group stems from the European bourgeoisie, which gives it power, as the socialist and communist groups stem from the European labor movement, which gives them power. This is an ideological and class background that is sufficiently clear to separate and distinguish the two groups. Yet they also are political groups, whose ultimate aim is to reach power or to make use of it. And the logic of power clearly shows in their conception of urbanization, in their priority choices, and in their methodology. This polemological aspect has allowed me to study and present them in a parallel way, as I think it is useful to do.

To sum up, I suggest that in the near future the following directions will give a wider audience to both doctrines in the making (which are not the only ones in Europe, as I have already said):

- the stoppage of the territorial growth of the urban area and the questions about the form of territorial expansionism;
- the research of a new urban unity and totality of the shattered European city: periphery-center relationships, historical web versus new web, essential functions, production and service sectors relationships;
- a redefinition of the relation of the city to its territory: rural-urban relationships, preservation of the rural area immediately

bordering the city, relationships of built-up spaces and open spaces, conservation of nature and of the built-up environment;

- a resurgence of phenomena of confrontation and negotiation between political and urbanistic programs, between the state and the dwellers: new means and practices of participation, influence of the association movement; and
- a redefintion of re-creating of social life units: units and relationships of neighborhoods, districts, sociality phenomena, human size, rejection of urban anonymity in a split-up time.

Both substitute urbanism doctrines in the making we have examined are new and creative in the conception of their priorities and methodology: they are of long standing and valuable on account of the political, scientific, and intellectual traditions in which they are rooted. I have selected them among others because these two urbanism doctrines are already quite well positioned in the political and cultural "adventure" of the Council of Europe, whose activity has been considerably intensified in the field of urbanism. And their influence will certainly be perceptible in the whole of Europe, as far as environment and social urbanism are concerned, yet also in matters of the future of the European industrial city.

*   *   *

To conclude, I should like to recall some of the important aspects of the recent rise of the theories and doctrines of a substitute urbanism I have examined. The fact that the intellectual and political forces that lie behind the present movement of a substitute urbanism are defenders of the past of the cities and of their historical cultural identity, considerably alters the data in a positive way. This is in relation to the vanguard trends of 1900-1930, which were elaborated on the basis of antihistoricist conceptions that were often simplistic and disadvantageous for the local situations and culture. But will those forces have the faith, the dynamism, and the optimism about the future that stimulated the vanguard of the 1930s?

One of my basic hypotheses deals with the rise of public opinion in urbanism, an urbanism that is oriented toward the future. I am concerned with the deeper expression of the movement of democratization which reappeared around the 1970s, after the disruption caused by Fascism and Stalinism in Europe, followed by the ascendancy of scientific and technocratic apolitical trends over the reconstruction of housing and cities. I leave out the negative aspects of that process—such as, for instance, the disorganization of university research. But I want to note the resurgence, in a new context and under original and positive forms, of the associative phenomena and movements that have deeply marked these trends.

In Europe, the recent process of repolitization of the urban and housing issues, clearly expressed by the methodology and doctrine of the neoliberal and socialist-communist trends, is a phenomenon deserving a more extensive investigation. That signifies not only a drastic readjustment of the function of the university researchers, but also an intensification of a municipal power that is not characterized by bureaucratic power, but by negotiated, comprehensive power. We can question ourselves: Would a closer connection between spatial planning and political organization contrive to avoid the trap of the 1900-1930s, when the unlimited trust in science and ideology directed the projects of urbanism toward productivism, which Berthold Brecht bitterly denounced in his poem, "700 Intellectuals Admire an Oil Tank"?

NOTES

1. The proposals for a substitute urbanism should clearly be distinguished from (and opposed to) the proposals of "overall projects for alternative urbanism"; the former are reformist proposals, new development models sprung from a criticism of the city crisis; the latter are overall alternative revolutionary proposals (such were, for instance, the socialist utopian projects). As far as I know, such proposals do not exist at the moment in Europe. Yet the Italian *urbanistica alternativa* experience may soon turn into overall projects for alternative urbanism.

2. Politization is here understood in its polemological, i.e., conflictual, sense. The French polemologist, Julien Freund, has defined polemology as the science of conflicts. See his work entitled *L'essence du politique* (2nd edition; 1978), Paris.

3. The concept of urbanization is here understood as the process of change in the modes of production and of the social and political uses of urban and rural space. This definition is close to, though not identical with, the one given by the Belgian sociologist, Jean Remy: His concept of urbanization is defined as "a process of structural change in

opportunities of practice." Jean Remy and Liliane Voye, *La ville et l'urbanisation* (Duculot, 1974), 65-66.

4. There exists an important tendency toward a critical reexamination both of the turn of the century in Europe and of the great authors who dealt with that period. See M. Cacciari, *Metropolis: saggi sulla grande città de Sombart, Exdell, Scheffer e Simmel* (Roma, 1973); L. Benevolo, *Storia della Città* (Roma, 1976); Y. Grafmeyer and I. Joseph, *L'Ecole de Chicago, Naissance de l'écologié urbaine*, (Paris, 1979); texts by Park, Burgess, McKenzie, Wirth, and Simmel; G. Piccinato, *La construzione dell' urbanistica: Germania 1859-1914* (Roma, 1974); see also S. Jonas, "Du quartier au voisinage: doctrines d'urbanisme de rechange et conception d'unité de vie sociale dans l'Europe de 1900," an article published in the international magazine *Architecture d'aujourd'hui*, No. 203, Paris, 1979.

5. Giovanni Botero, *Traité de la grandeur des villes*, 1606.

6. European Conference of the Ministers in Charge of National Development, Vienna, Austria, October 5-7, 1978. Final resolutions, Council of Europe, Cermat, Strasbourg, 1978. Let us recall the main directions we should keep in mind for the campaign for urban renewal, that General Secretary of the Council of Europe, Mr. Kahn-Ackermann, mentioned in his speech in Munich: improvement of urbanism, of sociocultural facilities, and of the communication, and transportation system; measures aiming at creating employment mainly for the local dwellers; action for raising a feeling of belonging to one community; launching of participation programs (Third Confrontation of Historical Cities, Muenchen-Landshut, Bavaria, Nov. 29-Dec. 1, 1978, CONF/Hist. (78)19. The Division de l'aménagement du territoire, monuments et sites du conseil de l'Europe contacted me in 1978 for an expert report, which was followed by an agreement for research on this topic: "Etude sur les voies et moyens de contrôler les transformation du tissu social dans les quartiers historiques" (Study on the Ways and Means of Controlling the Change of the Social Web in Historical Districts). This work was to lead to a recommendation by the Council of Europe on the rights of the dwellers to remain in the historical districts of the European cities. In a chapter of that report, I have already sketched out a primary study on the new doctrines of alternative urbanism that have appeared in the two important member countries, France and Italy. (Members of the research group, scientific responsible members, were Julien Freund, sociologist and polemologist, and Stephan Jonas, sociologist and urbanist. Other members were Yves Ayrault, engineer; Pier Giorgio Gerosa, architect and urbanist; and André Kocher, sociologist. Université des Sciences Humaines de Strasbourg, 1978.)

7. Council of Europe, Confrontation IV of Berlin, April 26-29, 1976; *Les grandes villes européennes face au chagement; un avenir pour leur passé*? Council of Europe (Strasbourg, 1977).

8. Confrontation VI, Ferrara, Oct. 10-13, 1978, *Vitalité des ensembles historiques: facteur et produit de l'équilibre ville-campagne*, Conclusions, CDAT, GT 2 (78)10. The Italian architect G. Campos Venuti used the right words in his talk to express the change of the Council of Europe policy toward cities: "From the original monument which must be restored isolately, to the web of the environment which must be preserved and cleansed as a whole, to the most recent conceptions of *unitary social and architectural protection,* the conception has progressively changed from particular to general, to the point where it now is an essential 'part of' but by no means separated from the urbanistic 'whole.'" (in "La conservation du patrimoine architectural: facteur d'incitation et de diversification de l'activité économique aux niveaux local et régional," Colloque de Ferrare, op. cit.).

9. We, of course, do not want to express a judgment of value about local and national actions and experiences, which will certainly prevail for years; their primary

doctrinal starting point will remain in the will to improve the pattern of the traditional and expanding urban growth, possibly by providing alterations—maybe critical ones— through a partial or punctual experimental process.

10. The Italian phrase *urbanistica alternativa* used by the authors is here understood as referring to a reformist project. Therefore I shall use it in the meaning of the French *urbanisme de rechange* (substitute urbanism).

11. "In positive words, *another way* should be chosen at the crossroads: ... to create new forms of life, of urbanization, of leisure, of culture. To give a qualitative dimension to the economic growth in order to turn it into *a 'soft' growth"* (Valéry Giscard d'Estaing, *Démocratie Française* (Paris, 1978), 17. "The 'soft' rehabilitation of older neighborhoods will be promoted by new aids from the urban development fund" (V. Giscard d'Estaing, Closing Address, p. 4, Premières Rencontres du Cadre de Vie, Palais de l'UNESCO, Paris, Dec. 5-7, 1977). "Today I present the French people with the Charter of Quality of Life. It marks a new stage in the struggle I am determined to wage with all men and women to concretely improve everyday life. This Charter is a social charter of everyday life. Its ambition is to reduce *ecological inequalities* and to create a happy environment for everyone" (my emphasis), in Charts de la Qualité de la Vie, Introduction by V. Giscard d'Estaing, p. 5, Service d'information et de diffusion, Ministère de la culture et de l'environnement, Paris, Feb. 1978.

12. Beside the various specialized publications of the DATAR and the Commissariat Général au Plan, see P. MASSE, *Le plan ou l'anti-hasard,* (Paris, 1965); *Réflexions pour 1985.* La Documentation Française (Paris, 1964); Le Groupe d'Etudes Prospectives du VI. Plan, *1985, la France face au choc du future,* (Paris, 1972); *Essais prospectifs de la France en l'an 2000,* La Documentation Française (Paris, 1973), research ordered by the DATAR).

13. S. Nora, B. Eveno, *L'amélioration de l'habitat ancien,* La Documentation Française (Paris, 1975); Union Nationale des Organismes H.L.M., *Livre Blanc des H.L.M.,* Habitations à Loyer Modéré (Paris, 1975); R. Lyon, *La situation actuelle et la politique de l'habitat depuis un quart de siècle,* Conseil Economique et Social (Paris, 1975); R. Barre, *Réforme du financement du logement,* La Documentation Française (Paris, 1976); M. LaBrousse, *Rapport sur l'aménagement du temps,* La Documentation Française (Paris, 1976), Collection "Environment," directed by Serge Antoine.

14. European Conference of the Ministers in Charge of the National Development, Bari, Italy, Oct. 21-23, 1976, "Urbanisation et Aménagement du territoire," Council of Europe, Strasbourg, CEMAT (76)2. See also the Conference of Vienna (op. cit.) in 1978, where the ministers "were satisfied to note down" the works and proposals of the "Recherche Prospective" expert group on "national development and prospective research." In these studies, the French research works have played an important part.

15. "Will this Charter consecrate a French way in environment policy? I shall answer yes and explain my opinion in front of a European audience. A 'French-like environment' in France is not synonymous of an inward retirement. It simply means that quality of life is tied to the ground. The ecologists would say that it is positioned and rooted. Quality of life must fit in the European contrasted and attractive geography" (V. Giscard d'Estaing, Closing Address, Premiéres Rencontres Européennes, op. cit., 5).

16. Chalendar Report, La Documentation Française (Paris, 1972); *Rapport sur l'aménagement du temps* (Rapport LaBrousse, op. cit.). On futurology and prospective methodology, see also B. de Jouvenel, *L'art de la conjecture* (Paris, 1964); J. Fourastie, *Les 40,000 heures,* (Paris, 1965); J. DuMazedier, *Vers une civilisation du loisir?* (Paris, 1962); and a more recent work, Abraham Moles and E. Rohmer, *Théorie des actes; vers une écologie des actions* (Paris, 1978).

17. "It must be acknowledged that Europe has sometimes smashed the mirror of its history by letting its city centers or suburbs get invaded by an American-like giantism or anonymity" (Closing Address, Premières Rencontres Européennes, op. cit., 2).

18. Giuseppe Campos Venuti, *Una esperanza di urbanistica alternativa: il piano regolatore di Pavia,* Italian contribution to the Housing World Congress organized by the United Nations in Vancouver, Canada, published in the Italian magazine of architecture Parametro. See also Franco Ferrarotti, *Una Sociologia Alternativa* (Bari, 1972).

19. Second Confrontation of Historical Cities, Bologna, Oct. 22-27, 1974, *Coût social de la conservation intégrée des centre historiques,* Council of Europe, Strasbourg, 1975; P.L. Cervellati and R. Scannavini, *Bologna: politica e metodologia del restauro nei centri storici* (Bologna, 1973).

20. See also G.H. Bailly, *La patrimoine architectural: les pouvoirs locaux et la politique de conservation intégrée,* Council of Europe, Delta S.A., Vevey, Switzerland, 1975; M. Jaeggi, R. Muller, and S. Schmid, *Das rote Bologna* (Zurich, 1976).

21. L. Benevolo, *Conservation et urbanisme; une esquisse de bilan,* in Confrontation of Berlin, op. cit., 9, emphasis mine.

22. Elio Veltri, *Urbanisme et Participation,* 17 pp., Premières Rencontres Européennes du Cadre de Vie, Paris, op. cit. text handed over in French by the organizers); G. Campos Venuti, *Une expérience d'urbanisme de rechange; le plan régulateur de Pavie,* op. cit. idem (this is the text translated into French and presented in Vancouver in 1976); E. Veltri, *L'unité urbaine: le centre et la ville: le cas de la ville de Pavie,* report presented by Veltri at the Premières Rencontres, op. cit. 1-7.

23. See in particular Theme II, Prof. Italo Insolera, *La conservation du patrimoine architectural: facteur de la reconquête de l'identitité culturelle et de la revitalisation régionales*: Theme III, Prof. Ghetti, *Législation et participation: moyens complémentaires d'une politique de conservation intégrée*; General theme: E. Detti, *Vitalité des ensembles historiques: facteur et produit de l'équilibre entre ville et campagne*; Carlos Cesari, *Le plan de l'aménagement général de Ferrare et la programme de recouvrement et revalorisation du centre historique,* Confrontation of Ferrara, Council of Europe, op. cit.

24. See in particular the studies and lectures of Prof. Lucius Burckhardt, President of the Deutsche Werkbund, at the University of Kassel: "Demokratie im Staedtebau: Pavia," in *Basler Magazin;* "Die Werte einer Stadt zueruckgewinnen," interview of Elio Veltri, Mayor of Pavia; Stephan Jonas, *Nouvelles théories d'urbanisme de rechange: quelques modèles italiens* (Strasbourg, 1978).

25. Federico Oliva and G. Campos Venuti, *Urbanistica alternativa a Pavia,* Marsilio Editori (Venezia, 1978); see also Campos Venuti, Confrontation of Ferrara, op. cit.

26. "This urbanist alternative for the cities has slowly been implemented in Italy thanks to the theoretical elaboration and the concrete experiences that the bravest city administrations have confronted in the most heterogeneous sectors.... These experiences were almost never carried out on a unitary level; on the opposite, the majority of Italian municipalities carried out very few to none of them. And yet, altogether, these experiences form a whole wide frame that can produce what I called the urbanist alternative for cities (Campos Venuti, Confrontation of Ferrara, op. cit).

27. During a conference for the defense of German workers' cities organized in 1976 by the socialist municipality of the city of Gelsenkirchen in the Ruhr, where I submitted a paper on workers' cities, an engineer from Darmstadt University, Dr. Martin Eisenle, gave a talk on his research about the city of Essen, in which he revealed that over the last fifteen years more old workers' housing has been destroyed in that city for reasons of land speculation and development policy, than during the massive destructions of

World War II. (Conference "Erhaltung von Arbeiter Siedlungen," Gelsenkirchen, 1976).

28. A systematic study on relations between political events and urbanism is still to be done. We are held back to indexes and hypotheses that only a future research will be able to confirm" (*Aux sources de l'urbanisme moderne*, op. cit., 139); see also S. Jonas, "Organisation et transformation des villes," an article on Benevolo, in the French publication, *Cahiers universitaires de la recherche urbaine*, No. 5, Paris, 1979, Editions CRU, 63-67.; J. Freund, *Le nouvel âge; éléments pour la théorie de la démocratie et de la paix* (Paris, 1970).

29. *Modi e forme di intervento della cooperazione nel centro storico: ipotesi e programmi*, published by the Nazionale Cooperative e Mutue Federazione di Ferrara (LEGA), Associazione Cooperative di Abitazione di Ferrara (A.C.Ab.), Consorzio Interprovinciale Cooperative di Abitazione (C.I.C.Ab.) (Ferrara, 1978); G. Ricci-Garotti and A. Cossarini, *La cooperazione—storia e prospettive* (Bologna, 1974); "La Cooperazione bolognese negli anni 70," in *Il movimento cooperativo*, No. 7 (Bologna, 1972).

30. We can sketchily summarize the five Rochdale principles: management and democratic control, free membership, distribution of profit, interest payment to members and inalienability of reserve funds. The French sociologist A. Meister, an expert on associations, defined the cooperative criteria as follows: open-door, democratic management, overcharge distribution (refund), and limited capital interest. *La participation dans les association*, (Paris, 1974), 24-25.

31. A Meister notices about associations, during a conference on urbanism organized in 1976 by the Ministère de l'Equipement, "I wish to keep away from the protagonists of participation who see in the associations a superior and renewed form of democracy, liable to rejuvenate the traditional representative institutions. Nothing allows us to say that associations are more (or less) democratic than municipal or regional councils or other representative institutions." *Participation et urbanisation* (Paris, 1976), 22.

32. "There does not exist any human practice (praxis) that would not tend to regulate future on the ground of experiences taken from the present." György Lukacs, *Sur la futurologie* (Budapest, 1969).

33. "Finally, the leading of long-run development raises the problem of some *control of science* for humankind and for scientifically advanced nations in particular. The idea of controlling scientists' works strikes our conception of science as a free disinterested activity by definition. Yet such an orientation probably meets a social need." V. Giscard d'Estaing, *Démocratie Française*, op. cit., 148.

34. Gaston Berger, *Etapes de la prospective*, (Paris, 1967), 286. See also "Sciences humaines et prévision," in *Revue des deux mondes*, No. 3, 1957; "Méthodes et résultats," in *Prospective*, No. 4 (Paris, 1959).

35. As early as the 1950s, the French—interministerial, ministerial, or other—state organisms had their own prospectivist structures, such as, for instance, the Groupes d'études prospectives of the various projects, or later on, the Système d'Etudes pour un schéma d'aménagement de la France (SESAME), a prospective cell of the DATAR. But it would be vain to deny the influence that the great international private groups of research on the future have on those prospective organisms. I should like to stress the particular influence that the scientists of the Association internationale des FUTURIBLES, and of the first generation of creators such as Gaston Berger, Bertrand de Jouvenel, Pierre Masse, Jean Fourastie, and Serge Antoine, had after 1968 on the

state urbanistic projects in France. That influence in particular allowed—at least on the whole, and in spite, of the great offensive of the American prognosticators of 1970-1974—the preservation of a certain importance of *social science* in the French urban prospective. See *FUTURIBLES, Théories et méthodes de la prospective: conférences à l'Ecole Normale Supérieure,* Numéro hors série (Paris, 1977), 107pp.

9

COMMENTS

Seymour J. Mandelbaum:

The two chapters preceding this comment are sharply different in their language, in their use of history, and in their attitudes toward markets and political hierarchies. My first response on reading them was that there was virtually no hope for anything but a polite but ritual discussion between the authors. I am, however, a perennial (and often disappointed) optimist in the face of communication barriers. These comments are intended to assess the barriers in the hope that there may be ways of overcoming them.

I propose to begin with Brian Berry's chapter—forcing him to carry the major burden of my assessment—because I feel more at home with his materials and style than I do with the issues and code words of European urban policy.

The Berry chapter has two beginnings. Neither is absolutely essential to the substantive argument of the essays, but both define the political position from which he consistently examines policy choices.

The first beginning might—wrongly, I think—be dismissed as merely a rhetorical device gone awry. The "American dilemma" described by Gunnar Myrdal was the conflict between an egalitarian ideology and racist practice. *Dilemma* is a long book, and my cursory reexamination may have missed an important element. As far as I can tell, however, Myrdal in 1944 did not prefigure his later critique of the progressive assumption of increasing international regional equality. That critique (articulated in 1956[1] and repeated often thereafter) has not been addressed to regional differences in the United States but to the gap between poor lands and rich. The transition at the beginning of Professor Berry's second paragraph—"the kind of central government controls sought by Myrdal never came"—poses a false contrast between reality and a prescription never rendered. It does, however, point directly to what in that paragraph (and writ large, in the corpus of his work) is the normative core of Professor Berry's view of the relations between urban policy and social choice.

I describe the core as "normative" intentionally, but not without some hesitation. The explicit form of the paragraph— the second beginning of the chapter—is positive, not normative, a description of what is, not what should be. The terms of the empirical argument are, however, drawn directly from normative theories of markets, market failures, and democratic politics. Professor Berry might have argued that American urban policy is patterened *as if* dictated by the premise and beliefs described in the second paragraph. He has, however, chosen a very strong form of empirical assertion which is highly vulnerable to counterexamples. American urban politics is riddled with efforts to inhibit the movement of capital and labor, and to correct properly functioning rather than failed markets. The expansion of the housing domain to include a broad range of services and neighborhood environments, and a pervasive concern with both natural environments and with the equity of all public services, have combined to extend the range of national (not merely public) policy over the entire urban field. The concept of an essential minimum is still sometimes employed as a legitimating mantle, but it has been stretched over housing and community development programs, to which it barely applies.

The present form of the second paragraph does not affect the substance of Professor Berry's principal argument (beginning with the second section) but it constricts its political implications and the possibilities of comparative analysis. The argument, as I understand it, is that the character of small areas within central cities in the United States is strongly influenced by intrametropolitan, interregional, and even international relations. Though neighborhood redevelopment may attract a great deal of interest and publicity, the opportunities for local change will continue to be severely limited in the absence of shifts in the larger contexts. The contextual influences that have shaped the character of central areas are, however, powerful and of long standing and will not be easily altered. If cities sometimes seem the creation of a hostile foe, we must recognize that the "enemy is us." Even vigorous advocates of local change struggle with the tugs of cherished but conflicting values of conservation and of dynamic growth. (Professor Berry, quite self-consciously, recognizes the two strains within himself. He simultaneously wishes to soften the transition costs of social change and to accelerate its rate.)

History—as a record of secular change—becomes for Brian Berry a form of cultural destiny which only may be overcome by "enlightened leadership" responsive to a major crisis. The cultural analysis depends on the empirical descriptions of the social commitment to markets and of the complex pluralism of the U.S. political structure.

If you treat those descriptions as problematic—even for a brief, skeptical moment—then the conclusion of the chapter seems remarkably abrupt. Suppose, in that moment, that you imagined that the second paragraph described particular policies and political coalitions—that it leaped outside the United States to deal with multinational corporations, trade blocs, and the political economy of the international system of settlements. The conclusion would then (if consistency ruled) focus on policy changes, strategic shifts, and countercoalitions. The conclusion might be pessimistic, but it would not depend fatalistically on crises and leaders.

Three salutary influences would flow from a shift in the argument and form of the (now much abused) second paragraph. The first two are easily described; the last is more complex and requires some explanation.

As to the first: Professor Berry's treatment of large secular trends is matched by a gross dichotomy between obsolescence and (what I assume might be described as) modernity. A finer grain in the history and analysis might reveal small but important policy opportunities. Were, for example, the rapidly changing St. Louis neighborhoods described by the Washington University group[2] obsolete and doomed by cultural preferences and inevitable aging, or was their history shaped by political decisions that were and continue to be vulnerable to change?

The second influence addresses the purpose of this conference. Professor Berry is a superb comparative analyst. In the present essay, however, the potential comparative dimensions (e.g. homeownership, migration, decentralization) are not directed to the interaction and learning that has characterized the relations between European and American urban professionals for more than a century.[3] His analysis, like that of the romantic cultural historians, allows for comparison but not the transfer of traits (or technologies, programs, and political savvy). Only at a finer and somehow less historicist grain is learning possible.

The third influence (which I hope will be peculiarly appealing to Professor Berry) relates to a confession of ignorance. I would find it very difficult (or, more honestly, impossible) to write a final paragraph (or chapter) of the sort I have proposed. President Carter's Urban and Regional Policy Group certainly seems to me to have failed at the task. It may be that Professor Berry might succeed if he put his mind to the work but, I suspect, he is (with a considerable political passion) convinced of his own incompetence. Let me explain, lest a compliment appear an insult.

Professor Berry quotes extensively from Title VII of the Housing and Urban Development Act of 1970. For a time, at the end of the 1960s, members of the U.S. urban policy community were fascinated with the idea of controlling the distribution of population so as to create what might be imagined as the "well-designed" nation. Underlying this fascination was a traditional critique of rural "decay," urban "malaise," and suburban "sprawl." The idea of redesigning the nation to meet this profound set of ills was infused with a combination of an architectural vision and a regional science—including Professor Berry's own work on growth poles—enlarged beyond its own limits.

The fascination was short-lived.[4] Even before the emergence of the current regional debate and concern about the decline of largest cities, it was killed (or at least critically wounded) in the process of writing both the first biannual urban policy report and in the preparation of the Rockefeller Commission recommendations on population growth. American urban scholars are generous with their policy recommendations and are quick to wish that particular cities or the national government would articulate and act upon a coherent urban strategy. At the same time, most of them are, I believe, thoroughly skeptical of the idea of the authoritative spatial allocation of people and functions. This skepticism is certainly fed by a generalized appreciation of the uses of markets and the limits of hierarchical organization. It is also rooted, however, in a recognition of our specific ignorance. We are not prepared to specify with any certainty the implications of major alternative designs. We are more confident of our ability to discern errors and of the importance of error-correcting processes than of our knowledge of how to act correctly.

A confession of ignorance at the end of Professor Berry's chapter would offer, in my eyes, even more intriguing dialogic possibilities than any specific political program that he might present. (It would also, as an aside, soften the immobilizing force of his pessimism on the local activists who struggle to improve particular neighborhoods.) Why is it—at least as viewed at a distance—that Europeans seem so confident of their knowledge and their ability to control urban development? How are knowledge and control related for them? How is it that they are so optimistic about public planning and so deeply critical of "technocracy"? Is there a discrepancy between their frequently professed respect for American urban and regional studies and their freedom from our bug of doubt?

There must be two parties to a dialogue. If Professor Berry presents a more inviting face to the East (and perhaps it is also to the Left), Professor Jonas will have to reciprocate with perhaps an even more strenuous effort at reorientation. My suggestions for such reorientation are, however, broached very tentatively. I sense that I have read an essay in a language I overtly understand but in which the context in which words have meaning is almost entirely foreign.

My first suggestion is that Professor Jonas recognize that perception of difference. My antennae were alerted at the very beginning of the essay. He argues at the beginning that urban

growth has been "wild, expansive and ... unregulated," and
that cities have tended to "dissolution." In the U.S. community of urban scholars, the use of these terms associates him
with a vision which is not merely old-fashioned or conservative
(he may delight in such designations), but which is insensitive
and uncomprehending. The great force of urban scholarship
has demonstrated that growth has been expansive, but it has
certainly not been "wild" or "unregulated"; cities have
changed, but they have by no means dissolved.

I assume that the words on the first page and the others that
follow (artificial versus natural, gigantism versus human size,
hierarchy versus interdependence) are not part of that severe
international lexicon that has been stripped (at least overtly) of
affective meanings. Instead, they are symbols of affiliation and
weapons in a political conflict. Professor Jonas should understand—if I am right about his intentions—that the terms are
an obstacle to understanding the position they are designed to
serve.

The chapter includes long lists drawn from the various formulations of an urban alternative. Let me risk a distillation of the
various urban strategies which interest Professor Jonas. The
alternative strategies often (though not uniformly) include

(1) restrictions on deconcentration so as to preserve the centrality
    of established city cores and to avoid the social and ecological
    costs of discontinuous development;
(2) restrictions on the total size of the largest cities and a com-
    plementary effort to revitalize small and medium-sized in-
    dustrial and commercial towns;
(3) the preservation or expansion of manufacturing districts near
    the core of large cities, avoiding thereby the shift in residential
    neighborhoods that follows the displacement of manufacturing
    by information-handling functions; and
(4) agricultural, trade and communication policies that restore
    city-hinterland relationships to what they were at some point in
    the past.

I recognize that (even if it is formally correct) mine is a flat
representation of the urban alternatives. The flatness has,
however, one obvious advantage, and, I hope, Professor Jonas

will correct rather than reject it out of hand. The long programmatic lists virtually suffocate all aggregate and distributional questions; the flat version highlights them. Who among the great variety of producers and consumers will be advantaged by the spatial restrictions and who will be harmed? Will total production be higher or lower? Will it be of better or of worse quality? Are we currently underinvesting in housing in relation to other economic goods, or overinvesting? Who will be granted an entitlement based on long residence, and whose liberty to move will be curtailed? Which farmers—in France and in Iowa—will be hurt to aid some others in central Italy? Who will be delighted and who will be kept in ignorance if communication is restricted?

In order to address these questions, Professor Jonas will have to expand the manner in which he employs historical arguments. The chapter treats the past either as a "patrimony" to be protected against abuse or as a condition to be remedied. Neither use will support the modeling enterprise required to address the issues posed by the urban alternatives.

If Professor Jonas shifts linguistic codes and clarifies the implications of his alternatives, he will open to the West (and perhaps to the Right) so as to strengthen an international discussion of urban policy. Failing this and at the worst, these comments may encourage communication across the Atlantic by interposing a common annoyance between two authors.

NOTES

1. *Development and Under-Development: A Note on National and International Economic Inequality* (Cairo, 1956); and *An International Economy: Problems and Prospects* (New York, 1956).

2. James T. Little et al. *The Contemporary Neighborhood Succession Process: Lessons in the Dynamic of Decay from the St. Louis Experience* (Institute for Urban and Regional Studies, Washington University, St. Louis, Missouri, 1975).

3. For a recent example, see Ann L. Strong, *Land Banking: European Reality, American Prospect* (Baltimore, 1979).

4. I have dealt with this issue at some length in "Urban Pasts and Urban Policies," a paper originally prepared for the Russell Sage Conference on History and Social Policy, October 21-23, 1977.

John B. Sharpless:

Rather than dealing specifically with the chapters above, I would prefer to share some of my general reflections as I read through them. These ideas have always interested me as I've done my research.

The first idea has to do with our perceptual bias about the term "the city." I think this perceptual bias is so deep-seated in us that we have a great deal of difficulty overcoming it. The city is a pure artifact of human creation; because it is a direct reflection of our collective psyche, we all have a kind of antagonism as well as an attraction to this phenomenon we call the city. That bias affects not only our vision of urban futures but our selectivity of our city histories and also the map we make of the urban past through the present and into the future. It shows itself in a number of areas. It affects the judgments we make as individuals as we write our papers on what constitutes the orderly and the disorderly. It also affects our choice of whether or not to use a psychological, individual approach to the city or an ecological or systems approach.

This bias also shapes our own sense of nostalgia about a city that at some time in the past was presumably more humane than in the present. We use words like "a city of human size," as if somehow, way back in the Middle Ages, there was a city that was much more comfortable to its inhabitants—that the dung which was knee-deep in the streets did not smell and that people were happy in their condition. It also has to do with our cry, "urban crisis." One is struck by how often observers of the city are concerned with the impending crisis of the city—one can find this concern in the seventeenth, eighteenth, and nineteenth centuries, and now in the mid-twentieth century. It is one of those phrases much like the one my father used to say to me—that the youth these days are just no good. I think Plato said the same thing. I am struck by that, not as a criticism of these chapters, but because I too have a vision of a city that probably never existed. My city is not filled with people, it is filled with computer printout. My city is a systematic interaction of components. I don't talk about people; I talk about theoretical frameworks.

The next point: In the Berry chapter there is a discussion of what some people call the gentrification of the city, or what Brian Berry calls the private-market revitalization. I think it is interesting that some people use "gentrification" and some people use "private-market revitalization." But one of the most frightening and awesome implications about that discussion is that only a very few cities and only a very few neighborhoods will undergo this renaissance, which consists of throwing out the black and the Puerto Rican, redoing the households into $250,000 condominiums, redoing the basements so they can hold a Mercedes, replanting trees along the promenade, and making sure that trendy people in trendy clothes have animated conversations in shops on the corner.

Unfortunately, the rest of urban America is going to miss out on all the fun. What is most striking is that it is unlikely that these spas of cultural development are going to pop up in Gary, Akron, Cleveland, or Newark. And as Brian Berry points out, this may have something to do with the ratio of current housing stock to future housing stock. It may have something to do with the availability of those service jobs that draw people who want the fancy brownstones renovated, but it seems to me it also says something about a historical tradition in Berry's cities which a professional service elite established long ago. Washington, New York, and Boston have always had a professional service elite. In such cities as Gary, Newark, and Akron, the professional service elite was transitory and the owners of the factories sent their sons and daughters east to prep schools; their sons and daughters, perhaps with a great deal of insight into the future of Akron, stayed in the East. Thus, these cities cannot draw on that tradition of revitalization which I would say has gone on time and again in other cities throughout history.

Now, what of the cities that are left behind? To this point their histories have been rather depressing for them, and I think this is going to continue. They will continue to be our forgotten cities. One of the most awesome spectacles to me is Newark. There is no hope in a real sense for Newark; it is one of those cities where, when you ask someone who was born there, "Where are you from?" he says "New Jersey." He is very careful to separate himself from his place of birth. Similar sorts of things go on with people who escape from Gary, Akron, and Toledo.

There's another question we must answer about the revitalized cities, one that in the 1980s may be the most serious one we face: Where did the people go who once lived in the brownstones? I would suggest they have gone to FHA suburbs. The big issue of the 1980s may well be the suburban slum. It is something that I am sure the Europeans in the audience are familiar with. There are many European cities that have long had suburban slums; but for Americans it will be something new, and I am sure an enterprising graduate student will make himself or herself a career in the 1980s pointing out that the origins lie much before the 1950s. He will point out that Watts, after all, was a suburban slum. He will go back into the nineteenth century and show that we have had suburban slums before and that we just forgot they were there because we were so concerned with our downtown slums.

Another question I have is what is going to happen to the successful cities of the 1970s, what is going to happen to the Houstons when their role in the economy has passed them by? Long ago, first with the writings of Geddes, later on, with the writings of Mumford, and subsequently with Eric Lampard's work and Brian Berry's work, there has been the talk of the evolutionary model of cities, that is, how cities interact with the larger economy—how they sometimes fit in, sometimes not.

There is a behavioral psychologist by the name of Silagman who, in his criticism of Skinner's model of behavioral response, pointed out that there seems to be something underneath, inside the animal, which determines whether or not it is going to be receptive to learning of a certain kind. Silagman says that simply at the level of rats and mazes and training chickens to scratch and that sort of thing, there seem to be three types of situations that give rise to the ability of an animal—and presumably human beings and cities, which are collectivities of human animals—to learn. There are three different states: One is that you are prepared to learn, you are ready to go as the situation changes. Then there are situations where you are unprepared but you are indifferent; that is, once you awaken to the situation, you can readjust and become prepared. Finally there are some situations where learning is impossible because the animal is counterprepared.

I am often struck by cities being in this third position over time. The shape of the economy changes and some cities fit right in. We think today of a service economy, and the cities that offer services find themselves in the very best position. In an age of high resource use and resource shortages, cities like Houston find themselves in a very good position. We are in an age of declining importance of industrial activities, relative to services; therefore the old industrial cities are increasingly counterprepared to handle modern market demands. The problem we face as urban historians, as urban planners, is how to know if we live in a city that is counter-prepared. If we have any sense of social conscience, we might want to let other people know that they are living in a community that is counterprepared, so they can leave too. It appears to me that as one looks over the history of cities, the plight of Gary is not new—that it will be transferred in future decades to the cities that have adapted quite well to a service economy (maybe the "disservice" economy). But I would anticipate that the Houstons of the world will have their day as well.

This has brought me to a question I have always had. I am always muddled by the concept of the service economy and how I am supposed to handle my urban studies in the context of the service economy when we have very little in the way of good, healthy, robust theory to help explain how service economies work. Conventional strategies and models from classical, Keynsian, and neoclassical schools offer limited insights into the dynamics of supply and demand in an urban economy. My favorite example is medical services: If there were a sale today at a large hospital on open-heart surgery, very few people in this room would avail themselves of this wonderful offer. On the other hand, if you have a cardiac arrest on the street, you do not go hospital shopping. In other words, it is a case of a perfectly inelastic supply curve. You do not lie there on the sidewalk when the ambulance comes up and say, "What do you charge?'' As it turns out, when the doctor holds your heart or your liver in his hands, he can ask for everything, and of course you promise everything as well.

I would argue that there is a whole range of public services very much like my cardiac arrest example, and that we do not have good models to handle them. I do not think conventional Marxist theories offer much help either; the labor-value theory can offer some models as well. Think of the following problem: The effectiveness of police, doctors, and fire departments is ultimately measured by their declining usefulness. Correct? Therefore their value as laborers in the economy should be the opposite of what we want from them. When we have a great deal of crime, we should pay our police very little; when we have very little crime, we should pay them a great deal. There are other areas where the service economy models are weak. They do not handle all sorts of questions about what constitutes resource allocation and transportation use, the role of information, and cost structures—present cost versus future cost. Obviously, cities are in many ways the accumulation of past cost. The old boys who built the cities are long since dead, and they did not put any money in the bank to clean up their mess. What will be the mess left by the service economy?

I would like to close by observing that the cities are never complete. They are always in the process of becoming and ending. By that very fact, growth automatically and simultaneously generates decay, and decay spawns growth. It seems to me this cannot be used as an excuse to forgive people who are cruel to one another in cities, but it is an observation that cannot be forgotten. To call for or to anticipate the end of that process of becoming or ending is to call for a heavenly city upon a hill, a city out of time and out of space, a city without a past and without a future, and, I would argue, a city without people. In that process there will be rubble left behind, and the tragedy is that the rubble will be people.

IV

THE SURVIVAL OF INDUSTRIAL CITIES

Each of the participants in this final section, Eric Lampard, Giorgio Piccinato, Charles Tilly, and Sam Bass Warner, Jr., may have gathered, in Tilly's terms, as a "badaud." However, singularly and collectively, they served as more than the lounger who observes what is going on and overhears things but does not participate very actively in them. Their remarks contributed greatly to understanding the dynamics of modern industrial cities. Not only did they help to summarize and synthesize the conference, but the four scholars stimulated further consideration of the important issues related to the survival of industrial cities and to the concerns of urban historians.

In the following discussion, Lampard considers the implication of the industrial city no longer being "the engine but part of the train." Industrial cities and urbanization may be something one might write an epilogue to, rather than a force that offers dynamism to society. In fact, for Lampard, "cities" may simply be a code word for broader economic, social, and political issues. Italian city planner Giorgio Piccinato asserts that the future of urban Europe, not unlike recent trends in the United States, rests with small cities rather than large industrial ones. If planners do not adjust to such transformations, their work will prove useless. Focusing on another dimension of urban life, Charles Tilly sees a need for more discussion of power within the urban industrial context. He recognizes the importance of the connection between decisions regarding work and other urban locational determinants, which are also shaped by state action. His discussion also aids us in understanding the differences in allocation of space in European and U.S. cities. Finally, Sam Bass Warner asks how urban historians might attempt to say something "useful" and thereby "try to use the past to criticize the present and get our courage up about the future."

For Warner, the urban historians and others who gathered to discuss the dynamics of modern industrial cities were not unlike the nineteenth-century housing reformers described by Lutz Niethammer. These historians and social scientists may not have agreed with this assessment and, in discussion following the panelists' remarks, they raised a number of pointed questions. Roger Lane of Haverford College pointed out that conference discussion displayed an absence of old-fashioned liberalism. Noting that if a generation ago we were told that a third of the nation would be living in public housing, it would have sounded marvelous, Lane asserted that such a sentiment seems to have disappeared. Warner countered that the political climate had moved to the right and that those present were part of the process. However, more significant was the fact that we are beginning to see the limitations of the admirable social democratic planning and housing construction that had occurred all over the Western world. "We have yet to decide ... what the politics should be to remedy those shortcomings. We're waiting for the next Messiah." Eric Lampard suggested what the politics were that led to the current situation. Liberalism in the United States had sanctioned the use of public resources to underpin private resources and to take the risk out of being a private entrepreneur. He argued, "They've done a marvelous job if you think of the vast amount of housing that is literally under the fair housing acts and is federally financed. They never did intend to provide public housing in the sense of publicly owned housing."

Corinne Gilb, an urban historian who headed Detroit's Planning Department, mentioned the effects of the great world shift in the dispersal of manufacturing and industrialism in general. Just as the north of England declined as an urban area with diminishing textile production, the United States, for example, was confronted with a similar trend in steel production. As other nations moved to "command economies," which subsidized their own steel manufacturing, this could not avoid affecting not only industrialization in America, but urbanization as well. Moreover, Gilb pointed to the omission of discussion of a phenomenon she called "uncoupling." The creation of single-person households in increasing proportions also affects the shape of America's industrial cities, as does the communications revolution centered on the computer. Alvin Toffler's prediction is that we are going back to the cottage industry because the computer allows decentralization. As a consequence of such change, all kinds of activities will be going on in the home rather than in great aggregated buildings like the nineteenth-century factory or the early twentieth-century office building. While Gilb did not carry forth her discussion to consider the implications for gentrification and the effects of energy limitations on U.S. cities, it would seem that the issues she raised point to at least continuing, if not increased, decentralization of American society.

Gilb voiced concern that the scholars present did not take into account relevant empirical evidence: "We are tending to look at the work around us through somewhat anachronistic and very circumscribed lenses." She voiced her frustration with "the role of the intellectual or the role of the academic in terms of planning for the future rather than just doing an ex post facto interpretation of the present. We're terribly poor predictors." Another participant, Charles Connerly, a Visiting Scholar from the Division of Policy Studies of the Department of Housing and Urban Development, raised an equally critical issue regarding the policy implications of the conference. Arguing that he was not convinced that history had a policy relevance, Connerly, a planner, claimed no true suggestions for change had been offered. The nature, role, and significance of urban history had been challenged.

The session, and conference, concluded with an impassioned statement (in French) by Stephan Jonas of the Institute of Urban and Regional Planning at the University of Strasbourg.

Jonas asked why industrial cities were being discussed. "I believe that the question posed ... is symptomatic, significant, and very, very important." He thought a moderate optimism pervaded the assembly, yet "survival" was a defensive term. In fact, survival, for Jonas, stood as the essence of urban reality. Survival implies responsibility; otherwise, one accepts Herman Hesse's position. When faced with the rise of Fascism, Hesse commented that everyone should look out for his own survival. Jonas thought such a solution inappropriate. Cooperative effort was necessary for the city, which, according to the Strasbourg professor, inherently had a greater possibility of growth and permanency than the agricultural town. While the latter has existed for a longer time and undergone several changes in production techniques and social organization, the city's potential for future survival is solid, despite the appearance that our cities seem sick.

Hence, the conferees both questioned and affirmed the survival of industrial cities; likewise, some attempted to use urban history as a policy science, while others inquired whether it had any policy relevance at all. Wide areas of agreement emerged alongside equally wide areas of disagreement. As industrial cities and the discipline of urban history develop during the remainder of the twentieth century, the questions, issues, and problems considered at the conference may be resolved. The survival of such cities and of our discipline await the determination of the future. Historians, however, must continue to look to the past for an understanding of how we arrived at the present.

10

COMMENTS

Eric E. Lampard:

The president of the University of Connecticut said to us at the outset of the conference that "the past is prologue." I have the feeling that with respect to "industrial cities"—a topic scarcely mentioned at this conference—that the present is already epilogue; that, as the detergent manufacturers' association said in response to the environmentalists' complaint, "soap had its chance 20 years ago and failed." The industrial cities had the nineteenth century and failed or succeeded, depending on your point of view. Anyhow, here *we* are, late in the twentieth century, and so many of our "urban" concerns perhaps are indicators of the extent to which we cannot really afford to be bothered with cities any more. It is not clear to me that cities still represent the kind of conjuncture they once did when they were able to yield those mysterious "external" economies of agglomeration and scale that helped give urban products a competitive edge over many of their rural counterparts. Of course, whether cities were ever such profitable arrangements as they appear depends on who is keeping the books. The bookkeepers were generally hired by the ones interested in progress and, both as business and national accountants, they tended to give us the impression that the goods outweighed the bads, that cities were here to stay and through them we could expect to enrich ourselves over a very long run.

I will not attempt to pull our disorderly proceedings together (or apart) but I feel obliged to say that none of our contributors has indicated what is meant by the adjective "industrial." No one does; "industry" is not a term that looms large in the technical vocabulary of economists. Certain others keep the notion alive; they have recourse to that comfortable hyphenate, "urban-industrial," and resort to such edifying noncategories as "preindustrial/postindustrial," whereby they indicate they do not know what they are talking about, referring to it as something which "it" is not—namely "industrial," a category not very clear to start with. It is not a happy situation. My notion of "industrial" will be that of increasing ratios of "capital" to "labor" and "land," as these terms are conventionally used and, more especially, the profitably substitution of forms of embodied capital for other factors of production in a given output, all within a system of differential possessory interests or property rights. Hence, our nemesis, the rising burden of fixed costs (not to mention fixed prices). Chronologically speaking, industrialization is an accelerated tendency in this direction, a fortuitous happening, a conjuncture of the late eighteenth and early nineteenth centuries first of all, booming on over into the twentieth, and still gathering momentum in other parts of the world as we move into the twenty-first century. It involves the peculiar occasion or circumstance of getting people out of the countrysides, where they and their forebears had always lived, and into "cities," and then finding something for them to do once they have gotten there, if they live so long. One might begin to think of cities nowadays—in highly urbanized societies—as being perhaps where countrysides were circa 1800. You are not going to get very rich keeping everybody in the countrysides, particularly if it is a much larger everybody than it was, say, circa 1700. Even in the United States, there was only so much people could do in "rural" settings, no matter how much of the continent they were able to rip off, with minerals or lumber to loot. You had to dream up new ways of "progressing"—"modernizing," as our social science friends tell us—whereby some people could bring together other people, energies, and materials in ways that could yield increasing returns. If not literally, at least the yield would be something substantially more "worthwhile" on average than a person could expect to get in his native or inherited type of social environment.

So the nineteenth century spends time and effort moving from rural to urban "idiocy." If some undiscovered locus of novel idiocy potentially exists, have no fear, we will find it, and, after a little bit of free collective bargaining en route, duly pronounce it even more worthwhile. One thing is definite: Increments of *urban* population have not been among the weightier terms in the economic equations of mid-twentieth-century high-income countries, even though they may have been, and perhaps still are, lively elements in the political equations. Most market systems, I would judge, notwithstanding Brian Berry's enthusiasm for "the private sector," have not seemed to get along very well since the 1920s without, shall we say, some blurring of the distinction between "private" and "whatever else" that Berry has reasserted so passionately. In the nineteenth century much profitably real-estate development was conditional on a growing commitment of local public funds by way of infrastructure, utilities, and services; the public had to invest in expanding its tax base. Since the Housing Act of 1934 a rising share of the underwriting and overhead costs of construction—from suburban housing, highways and airports, and tarted-up downtowns—has been billed to the federal taxpayer, and a lot of moneylenders, managements, unions, professional politicians, functionaries, and quite a few professors have done very well from it. Now there are a lot of empirical associations between the phenomena of getting people out of the countrysides and forming them into cities, and what has come to be called "economic growth"; and there are many other related tendencies and concomitants, which we might characterize as "social changes," that went along.

From the vantage point of economic history, the most important linkage perhaps is the fact that the pulses, and resulting momentum, of population concentration were closely geared to the "long swings" in rates of change of other key economic and demographic variables. Under nineteenth- and even early twentieth-century conditions, rates of increase in population, of major components of output (construction, producer durables, and so on), and of per-capita product alternately accelerated and decelerated over spans of from 18 to 25 years, peak to peak (as indicated by five-year moving averages for the United States). Periods in which ratios of urban to total population increased rapidly—peaking in the United States circa 1850, the early 1870s, the early 1890s, circa 1912, and 1926—coincided with peaks in migration of workers, manufacturing plants and equipment, buildings and public utility output, and in the supply of fixed capital generally. The surges of investment in construction, utilities, and urban households seem to have augmented or buttressed investment in producers' goods and the growth of capital stock in manufactures, and to have *followed* on an emergent labor shortage or state of full employment which served to induce further migration to cities. Commitment of resources to these extended, population concentration-sensitive, types of capital formation helped bolster spending over periods longer than short-period business cycles (characterized by fluctuations in the employment of *existing* facilities and workforce). Construction generally is the greatest single (albeit amorphous) category of capitalization in the nineteenth and early twentieth centuries, compared with which manufacturing, mining, electrification, or even automobilization are limited in magnitude. In fact, all of the latter in their respective investment heydays were tied into the phenomena of peopling and building the urbanized environment. While their impacts lasted, timed-lagged, urban-sensitive, spending on construction, utilities, and household formation had positive effects on total employment and output, and also affected the directions and timing of demographic responses (migration and fertility changes).

Another phenomenon is the rather close correlation, as Robert Higgs shows, between the rate of patenting inventions by populations in the various states and their levels of urbanization (particularly outside the South) in the period 1880 through 1920. Similarly, one can show significant correlations—using the data for states in Perloff et al., *Regions, Resources and Economic Growth*—between levels of urbanization and personal incomes per capita. Thus, one can simultaneously equate certified inventiveness with urbanization, urbanization with average income, and average income with inventiveness in the late nineteenth and early twentieth centuries (outside the southeastern states). There are clear portents of similar associations with respect to state levels of urbanization in the antebellum decades, although prewar data allow even less pseudo-rigor than do those for 1880, 1900, or 1920. From the 1920s such relationships attenuate and the "long swing" itself gets obscured by shocks from world wars and depression. Technical and other innovations, so essential to capital-deepening in the economy, no longer have the appearance of any association with incremental urbanization of population (itself a capital-widening effect). The next peacetime investment boom was part of the post-World War II adjustment experience, and was already falling off from 1949 until pulled over into the early 1950s by Korean War spending; it occurred in an institutional setting and under auspices remote from those obtaining before World War I.

One might view nineteenth-century urbanization as an almost unprecedented learning experience, in which tides of population are progressively brought into divergent kinds of social settings enhancing their average productivity and total output. Human interaction in these densely populated environments led again, I would think, under competitive conditions, to a serendipitous, casual consciousness-raising, mind-bending, and learning—a learning in the workplace, in the household, and in the surrounding neighborhood that I have elsewhere referred to as "the cell-like structure" of the city as an informal learning environment for mostly autodidacts. By the 1840s in the United States there was already a substantial income gain, on average, to be had in activities outside of farming and, to a lesser extent, in regions where income per worker reflected a greater commitment to nonagricultural pursuits. Now "agricultural" and "nonagricultural" are never synonymous or coterminous with "rural" and "urban"; yet, manifestly higher returns on average outside husbandry tend to reinforce and amplify structural departures in work and residence broadly associated with urbanization by contemporaries from the second quarter of the nineteenth century. The differential had certainly existed in some measure in many places before the eighteenth century, but it had nowhere launched a comparable ecological transformation.

These structural changes in work and residence are jargon for the discovery of new ways—cost-reducing, resource-saving ways—of doing old things and occasionally learning new, profitable, things to do. Either way, if one could so organize new ways or things and capture the fruits thereof, there was a capitalizable "opportunity." The quasi-rents or extraordinary profits of true inventors, unless protected by a patent monopoly, would soon be eroded by effective competitors down to some going "normal" rate. One could make money out of such protean environments when novel ways and things proved to be "worthwhile," proved to be part of the "advantage"; urbanization would have stopped or even reversed itself if, on balance, people had been able to find other, more rewarding ways of distributing themselves under nineteenth-century conditions.

Through the mid-1850s, later 1870s, middle to late 1890s, most of the 1930s, and, perhaps for different reasons, most recently, the "advantage" of a central-city location noticeably diminished, although circa 1970 there was still a discernible labor productivity gain, on average, with incremental size of metropolitan-area population. Much of the learning—and consequent locomotion of population—that went on during the nineteenth century involved little more than adapting successive migratory (and birth) cohorts to prevailing capital-widening urban conditions rather than adopting wholly original or novel models of behavior. These would be capital-deepening in their effect, insofar as embodied technological forms were substituted for what a later generation of econobabblers would dignify as "human" capital (instead of "hands"). The "advantage" of urbanized activities—mostly economies of specialization and scale—achieved by local agglomeration and/or more extensive and efficient transport and communications had to be sustained if people were to keep coming out of their native settings at home or abroad and forming into cities. This momentum was very "advantageous," in the economic accounting sense, so long as the incremental products of labor were positive. Thus, the peopling, building, and servicing of nineteenth-century cities—which did not take account of most social and ecological costs until they were converted into "problems" and capitalized by later generations—turned out to be an unexampled change in human ecosystems; emanating from the British Isles, it helped shape the size, content, and direction of aggregate activities in a growing number of countries. As a rising proportion of the enlarged U.S. population experienced the conditions of urban life and livelihood, an urbanized social system gradually *took place* in space and time across the continent.

Given the role of urbanization in fashioning the content of production and of concentration in sustaining migration, the momentum of the residential transformation was in this sense "a creator as well as a creation of 19th-century market society." In their very formative processes, the cities were powerful engines of economic growth. By the time (1910-1930) a substantial majority of the population had come to live in urbanized areas, however, further increments of city dwellers had a proportionately diminishing impact on their highly capitalized environments. By a variety of measures, there seems to have been a loss of urbanizing momentum since the 1880s (even earlier, of course, in the case of the United Kingdom). Thus, the earlier rates of increase in the urban share of total population do slacken over successive "long swings" in the wake of fertility decline, in the aftermath of reduced immigration, and in the gradual drying up of the historic reservoir of rural population.

This is not, of course, to say that fluctuations in fertility or migration have altogether disappeared. Alterations in the composition and distribution of population continue to have social and economic consequences locally and regionally, irrespective of the overall rate of growth. Nevertheless, to the extent that what I am saying about secular scalar tendencies is true (and that may not be very far), we are left with most of the working population living in and around urbanized areas. What I am doing here is trying to make something out of "industrial cities" to warrant all the attention that we have come to pay to urban history, to protect our own vested interest. Was population concentration, under nineteenth-century market auspices, merely incidental to the ecological transformation, or did it become an independent force—a vehicle pulled along with others in the train of economic growth or itself a powerful locomotive in the absence of which economic and social change would not have gone so far, let alone in this particular direction? If the making of industrial cities was the latter, then, as I suggested, much of the twentieth century has been epilogue; there have not been all that many people left out there to concentrate, at least within the political boundaries of the United States. If no more than people are needed to restore urban "vitality," well, there are Mexico, Haiti, Cambodia.

Obviously, economic change has not stopped; social change did not end with the low rate of population growth in recent decades; the metropolitanization of urbanized areas is far gone. Our accountants tell us that we have put on more real product in the past 30 years than in the previous 300. But in the United States, so-called development has been getting very little leverage from increments to urban population. To be sure, relatively underdeveloped parts of the country have been brought into closer parity with the rising national average in respect to levels of urbanization, metropolitanization, labor force composition, per-capita incomes, and so forth. But recent economic and social changes have probably not originated in such adjustments, although some politicians and their publicists—in the great tradition of "sectional" agitation—make capital out of the notion that "growth" of newer areas has occurred at the expense of older areas. Convergence among the regions has been going on steadily for a century and, apart from the special generosity of the military-industrial-congressional complex, much of recent capital-widening in the Southeast and Southwest may well indicate a failure of imagination in finding new things to do, other than working up available natural resources and "exporting" the going national average of ways and things to long-lagging regions. Alternatively, their more recent capitalization may mean that such areas now have concentrations of newer, more efficient labor- and resource-saving facilities (as well as cheaper workers and tax abatement carrots). Since the end of the long war-induced "hundred-month boom" in the late 1960s, and especially during the sharp recession during 1974-1975, the "growth" of the nation has been largely accomplished in states like Texas and California rather than in Pennsylvania, New York, or Ohio, where so much capital is tied up in obsolete or otherwise inflexible modes (not least in their antique urban infrastructure).

It is not that we lack imagination compared with our nineteenth-century forebears, but that the environment of industrial cities no longer generates the new things or even perhaps the new ways via the informal, cut-rate information exchanges that were once characteristic of larger urban centers (and almost nowhere else with the same frequency or variety). This is not to say that if you spend enough federal money you cannot make parts of Lowell look like the Left Bank, a haven for beautiful people; but real imagination does not come readily or cheaply in the course of transferring country remnants to urban work force. At ever higher levels of capitalization and fixed costs, "progress," so to speak, takes longer to realize and possibly yields less in the absence of needling price competition. Thus, urbanization in its later manifestation has not represented movement along an ever-descending learning cost curve. Learning is harder, narrower, and more expensive—it no longer "happens" in older, run-down urban environments and no one expects it to happen in shinier new models. Urbanization has long ceased to be a locomotive process, and judged by the selfless cries of mayors and their hired "consultants," cities are more "the problem" than "the solution," as the in-place policy scientists say after perusing their bottom lines.

Nowadays cognition itself must be capitalized, which will come as no surprise to etymologists. Unprecedented sums are invested (much of them directly, or at least underwritten, by taxpayers) during peacetime in R and D, specialized information-processing and communication networks, managerial and planning functions, and so on if the experienced environment is to be reinterpreted along profit-making lines. This "growth" imperative has required capitalizing aerospace "spinoffs," recombinant DNA; we may soon be down to our smallest silicon chips. Circa 1960 all the money was supposed to be tied up in nuclear power. Technology will save us from the thrall of diminishing returns. Thus, the system depends on consciousness-raising and brainstorming, not cities, for its dynamism; in the post-Sputnik era, for instance, it was necessary to dream up ways of getting to the moon before the Russians—money no object. Thank God-of-your-choice for the Russians. Much of our collective melodrama is endowed by media with this gamelike aura, as idiotic, I would suppose, as Europeans learning to smoke tobacco or drink chocolate in the sixteenth and seventeenth centuries. People had gotten by for centuries without such things back in Christendom (except for those who took off on the Crusades), but then they persuaded each other to abandon all that for the prospect of gaining the world. Think of the "growth," "wealth," and "progress" that were eventually created out of tobacco and raising cane before they culminated in lung cancers and dentures; other peoples' lands were seized, vast populations taken into slavery, and great empires really founded on smoke. At the time, the distinction between private and public sectors was conveniently fuzzy—until Adam Smith pointed out how counterproductive the erstwhile convenience had become.

As John Sharpless remarked, urbanization was a big artifact. The industrial city was itself an enormous production—eventually on a scale quite removed from earlier cities. Its use likewise entailed a great deal of nineteenth century imagination as well as inventiveness in social relations. Much of it simply involved furnishing an endless flow of country cousins with secondhand or second-rate urban equipage, but, as I suggested before, it also meant improvising novel combinations and permutations that were unknowable or unnecessary for earlier generations, in country or town: the capitalization of water and sewer systems, "savings" banks, streetcars, music halls, "news" agencies, the "quick" smoke, "fast" food, mental "health," you name it. But as the numbers or proportions of such cousins falls and the evaluation of their marginal urban performance drops, there's not much to be made employing them or equipping them for city life (except by social workers or politicians). To maintain corporate profits and pay union dues to those with a corner on the job markets, it takes more than a heap of living; you require a lot of auxiliary social organization to keep this mode of institutional accommodation going.

Even in the nineteenth century the industrial city was not just a private construction and rehab job. There was a growing volume of municipal indebtedness, public spending, and property tax. By the third quarter of the century one gets the first big crop of municipal bankruptcies in 1873, a bigger one in the mid-1890s, and a bonanza in the early 1930s when over a thousand "communities" defaulted on $2.6 billions, Detroit being the biggest. By 1937 there was a federal municipal bankruptcy act. One wonders if, even in that heyday of "privatism" (to use a copyrighted expression), the moneylenders and real-estate speculators could have financed local railroad connections, installed drainage systems or street lights, and "delivered" fire and police protection, as well as done their own things without some generous considerations passing to and from the political entrepreneurs who capitalized city hall. Even to undertake water or gas supply, run streetcars and eventually electrify them, and the like meant getting a franchise or leasehold on the city's public rights of way. Meanwhile, on the services to the needy side, the conscience-stricken and the busybodies pioneered most of their good-doing works—outside the almshouses and asylums—without much charge to the taxpayers, at least until the charities went "scientific" and "professional" around the turn of the century. But the enormous increase in local-government-bonded debt was already a matter for state legislating and federal census-taking before 1880. Public support of the private sector was critical if the urban artifact and its social spaces were to be further differentiated and extended, if real-estate values were to be maintained, and if the rent element in central land values was to be moderated by bringing accessible peripheral districts into the effective land market.

Public investment was necessary to maintain and enhance the accumulated stake in private urban property. Between the late 1850s and the early 1870s, the value of lots adjacent to New

York's Central Park is said to have increased by four times the average lot value citywide. Obviously, Henry George had caught on to the importance of public as well as private improvements by 1871; so had Simon Patten in his Jena dissertation (1878) on the *Finanzwesen* of North American states and cities, using Illinois data. The relation between municipal debt and extension of the built-up area was likewise featured in Henry C. Adams's *Public Debts* (1887) and Ely's *Taxation in American States and Cities* in the following year. Their views on municipal "corruption" differed somewhat from those of James Bryce and his reformer friends, who wrote off city governments as "the worst and the dimmest." It is, to say the least, ironic that the reformers' opinion of the foxy (folksy?) ward bosses "delivering" all those costly unstatutory services to the poor old uprooted has been confirmed, albeit in more exculpatory tones, by our latter-day ethnocultural romantics in their busy rewriting of "the shame of the cities." The economists, credulous as always, had meanwhile pointed out that nobody was spending all that much on the huddled masses, even in return for their votes. One still awaits some more empirical evidence on the matter as distinct from citation of oral tradition or self-serving memoirs of the Ed Flynn *You're the Boss* ilk. John Modell showed us that Philadelphians got some high schooling out of it, but the Philadelphia machine, for all its consummate corruptness, does not quite fit the roseate model of "community" control espoused by the current dissensus school of American historiography. Withal, the mismanaged public credit did furnish funds available from no other source. Every dollar spent by the public on capital improvements and protective services, as Frederick C. Howe pointed out in 1905, "adds its increment to the value of building sites." As he later put it, "The growth of the city is reflected in land values as in a mirror." There is still quite a lot to be found out about public investment, local politics, and city growth in the late nineteenth century.

The problem was (and is) to keep on finding something "worthwhile" to do on urban land that will prevent the peak central space from collapsing into the hole of a donut. How can central land uses be found quickly enough to preserve central land values or ground rents? Once upon a time, the "free market was supposed to do this; now it can no longer oblige, and city hall is on welfare. The receding perimeters—decentralization—already posed a threat to the city's core in the early twentieth century, even as it offered some relief from the costs of congestion and other density-related "problems." Put the other way around, why is it that property and land values cannot be made to depreciate *fast* enough to prevent the formation and spread of obsolete areas and blighted zones in the inner city? As Robert Murray Haig and other land economists realized in the 1920s, property values in blighted areas run down slowly, so that the costs of renovation always seem to preclude the possibility of their profitable current use. In a number of the older and larger cities, private enterprise could no longer afford the costs of clearance and redevelopment. Once the depression hit, the local taxpayer could not even afford to take over private property in order to help out.

The rising volume of municipal indebtedness in post-Civil War decades (when federal and state aggregate debts were falling) reflected the use of the public credit to support investment booms in railways and residential and nonresidential construction. In every major economic retardation down into the early twentieth century, the huge private outlay on residential, utility, and related infraconstruction overrode the ups and downs of business-inventory cycles and overshot the long-period surges in productive capacity. The kind of spending commitments made by or on behalf of people adjusting to town life and livelihood—without benefit of the long-term amortized mortgages or publicly financed mortgage insurance that only came in with the Feds after 1934—cushioned the effect of slumping "industrial" investments. Nevertheless, the marked secular fall in gross new construction (as distinct from repair and maintenance) as a share of gross capital formation, and, to a lesser extent, of GNP between the post-Civil War and post-World War II eras, was only moderated from just before World War I by the general rise in *public* construction expenditures. These took the peacetime form of highways, school-building, sewer and water systems, municipal plants and facilities, hospitals, and conservation and development works.

By the late 1960s, half the housing stock in the country had been built since World War II and more than three-quarters of the 70 million housing units in existence were, via one or another form of federal involvement, nominally subject to the terms of Title VIII of the Fair Housing Act of 1968.

With respect to the economic role of cities, my feeling is that part of the problem since the 1920s has been in getting a big enough peacetime volume of public spending to sustain the profitability of a system that no longer generated enough nineteenth-century-type steam. There was even a time around 1940 when Alvin Hansen and others thought that sufficient "compensatory spending" on the outmoded cities would help release the flows of savings necessary to heat up the stagnant investment channels. For the better part of a century, there had been that singular situation fostering what I have called the locomotive transfer of people and resources into concentrated and clustered modes of work and residence. It started with activities in which enterprisers were able to organize novel technological means in order to undersell existing markets; it spread to activities for which larger or more extensive markets could be promoted. The resulting "industrial cities," in American parlance, represented an "advantageous" ecosystem in which people and their effects multiplied. Nobody designed it; people learned by doing it piecemeal.

Some tried to design alternative systems; others caught on to Bentham's "cost-benefit" accounting rhetoric quite early and were able to show "advantageous" public as well as private improvements "in the public interest." So it unfolded; nobody envisioned the kind of outcome or the uncovenanted and unaccounted side effects of "progress"—least of all that war, depressions, and backlogs from both would obscure the "natural" rhythms and feedbacks that were assumed to make it all go on like some autonomous clockwork. Unless they were of an "historical" school, the policy scientists of the day failed to observe that incremental, cumulative learning by doing was *transformational,* not mechanical. Nor did anyone foresee that an urbanized republic would one day find itself confronting a grim choice (or nonchoice) of "growth or collapse," or so the economists tell us in their outrage at the eco-tinglers and nogrowers. By growth, of course, they mean more of the same unless and until somebody can come up with something more profitable to sell. Nowadays, such Tory cheek is almost indistinguishable in its implication from ritual Marxist chanting (in unison) about capitalism's imperative drive to expand!

Certainly, invention and capitalizing the social environment are no longer a matter of individuals and firms patenting or profiteering the urban scene. The city is not the engine but part of the train in which the poor are "delivered" services and the beautiful and gentrified people purchase "alternate lifestyles" (with a bit of 235 aid) in appropriate cars. Fundamental capital-deepening is carried on in some "nonplace urban realm" and seems to involve growing commitment of public funds, loans, and loan guarantees and a "socialization" of corporate and other private risk-taking. Metropolitan areas still house a growing share of the population; central cities are still a locus classicus for all types of specialties in manufactures, professions, and services. They are the home of the hype industries—advertising, promotions, moneylending and stock-gambling, public relations, press agentry, media, and a goodly share of higher education and the arts—while their armies of state and local functionaries have been duly delivered to the ranks of the labor monopolies otherwise shrinking from the decline of traditional urban manufactures. Even with hundreds of federal-state programs supposedly curing this or that urban ill, many municipalities maintained their solvency in the late 1970s by running down their sewers, water mains, streets, transportation, and other overhead on which the private sector depends. Since the partial recovery from the steep 1974-1975 recession, a rising share of city governments have experienced operating deficits of which New York's is only the most wanton and publicized. The fiscal situation of a majority of the 300 cities with 10,000 or more residents has been even more gravely afflicted since 1978 with reductions in many federal flows to which local governments and their clients had become addicted. So my feeling would be that "industrial cities" and urbanization are something to which you write an epilogue. In many parts of the country, cities are poor learning environments and the growing dependence of their public and private sectors on extralocal public subsidy in its various guises is itself a token of their metamorphosis.

Are we really concerned about "cities," or is that just a code words for perplexing economic and social, and hence political, issues with which we have the greatest difficulty coming to grips, and in regard to which cities may perhaps be encumbrances in terms of what the capital budget requirements would be to make them pay off? We dumped rural America, and it eventually cost a lot of money to wait for the farm folk to move out or die off.

Giorgio Piccinato:

Here, at the conference, we have two groups, the Europeans and the non-Europeans—or if you like, the Americans and the others. The Americans are mainly concerned with what happened and the Europeans are mainly concerned with why things happened. Both groups think that the other group is a little bit naive. I am trying to see what (both groups) have found out in common and what (ideas) we might be able to develop. There are two simple facts that appear in all the essays. One is that the factors that created the great urban concentrations no longer exist. At the beginning of the industrial period the economic advantage of low labor costs existed. This economic advantage was greater than the social costs—class tensions, low environmental qualities, and the like—but social costs today have a greater impact than they did at the beginning of the industrial period. The second fact is that the city has lost most of its value as an instrument to condition the conscience, because now we have the mass media that allows for an overall diffusion of a uniform urban lifestyle. The mass media have solved the problem of how to sell products directly to consumers: The city is no longer necessary to influence the consumers' market.

I think there is another point, which has become clear in the recent decade and was not clear at the beginning of the nineteenth century. The construction of the city has become the construction of the territory. Suburbanization and decentralization have become a gigantic source of profits. At the beginning of the nineteenth century, however, it was the construction of the big industrial city that was a gigantic source of profit: The city was not just something that happened to be there because of other activities. The construction of the city itself was a great force. Something similar happened in the twentieth century, and at a greatest degree after World War II. Public investment was very much involved in suburbanization in the United States, as it was in the diffusion of urban infrastructures throughout the territories in Europe and in this country. Suburbanization and diffusion of urban infrastructures have taken what was once the place of the large industrial city. Living conditions in the large, densely populated urban center are certainly different than they used to be, but not enough so. The relations between capital and labor and the subsequent level of exploitation that were possible in the first industrial cities are no longer possible today. Today, everybody has some degree of expectation. This degree of expectation can give rise to major conflicts, and is probably a key source of all the problems and tensions in the major cities, such as crime and drugs.

Decentralization not only of residences but also of production has, first of all, a social meaning, not just an economic meaning. Professor Berry's chapter and, even more so, Professor Jackson's chapter emphasized the function of public support for private property in housing, and this favored decentralization. Professor Berry also mentioned the trend toward smaller cities—historical centers or, in Britain, the new towns of the 1950s. Especially in Europe, these smaller cities, as opposed to the large industrial cities, are enjoying a great popularity. The population and jobs of these cities are growing. This indicates that European cities are not dying. I would be interested to know if this were true here also. The future lies in the trend toward smaller cities, not the abandonment of the city per se. The growing amount of public investment in all European urban centers is very much a reflection of this trend. Of course, the story of urban centers in Europe is quite different from that of America, but as Anthony Sutcliffe has pointed out, ten or twenty years ago we were all hoping that the American city would be the model city of the world. This was not to be the case. Things are going in a quite different way. I think that a comparison between the United States and Europe might be much more interesting today than it was twenty years ago, when everything seemed so clear.

As we all know, the great escape from the city was supported by public policy in housing. It was the result of a long, articulated series of attitudes, both economic and social, that were positive toward the new and negative toward the old. This escape from the city is simply a result of public policy's lack of a positive attitude toward the environment. That is, it is the result of private appropriation of the environment. Whenever public structures fail to give a decent environment, people try to create their own individual environments. If public policy fails to do anything for central cities, individuals will have to do something for their suburban houses. This is a chicken-and-egg problem.

The urban future requires that we, in the first place, rehabilitate urban centers, not just upgrade the housing stock. We must get back the dimension of social interchange. The sources of this private, individual escape are racial and economic segregation and homogenization. We must fight this trend if we are interested in the future of cities rather than just in single houses. As Professor Jonas pointed out, we need a collective reappropriation of the urban condition. We must return the urban condition to the people and not just divide it up for individuals. Industrial bourgeois cities are actually very private cities, much more private than they used to be. The achievement of a decent urban environment today requires the development of decent collective spaces, and this will require the changing of all sorts of attitudes. There would be enormous consequences in the decision-making process and in the planning and structuring of city government. The attempts being made all over the world to restructure city government have much to do with the attempt to overcome the century and a half of the private takeover of the city. These attempts will have enormous consequences in terms of planning principles and planning models. From my point of view as a city planner, there are great problems in altering planning principles and models that have been built around the fundamental goal of constructing a capitalist society. We had to deal with the location of industrial production, the reproduction of the labor force, and the formation and realization of the urban land rent. As Professor Lampard has pointed out, land speculation in European cities was an incredibly strong element in urban growth. The connection and the conflicts between industrial profits and urban land rent have been very important in urban growth.

City planning is now in a very bad position because its traditions have been built around a model of society that is, in many parts of the world, changing. This transformation of the social model has made city planning much more complicated than it used to be. In a transforming society, the traditional city-planning approach proves to be rather useless, and this explains, to a certain extent, the failure of city planning in the Western and the Eastern worlds. What we now must do is to establish new goals and to design the structural elements and instruments that can work in a transformed environment.

Charles Tilly:

In his sardonic book, *The Higher Learning in America*, Thorstein Veblen played an ironic game of looking for the analogy to the academic procession. He went through a list of possible metaphors, one of which was that of a religious ceremony. After rejecting this and a series of other analogies because they were not quite right, Veblen finally concluded that the closest equivalent was a circus parade. This analogy nicely reinforced his point: what the businessmen who were on the boards of trustees wanted their faculties to do for them. We can play the same kind of game with the role of the general commentator. One of the analogies open to us, appropriate to the urban context in particular, is that of the "badaud," that is, the lounger who observes what is going on and overhears things but does not participate very actively in them. Another analogy is that of the street preacher, who gathers a crowd and listens to the hecklers, or that of the pawnbroker, who takes used objects and puts them on a shelf. I do not have the nerve to be a street preacher or a pawnbroker, so I will have to settle for being a simple badaud. Let me tell something about what I have overheard in the conversation of the last two days. I've been "hanging out" in the "bounded unpredictability" of the street, in Martin Katzman's phrase; after I have given my comments, I shall slip unobtrusively back into the crowd.

The idea of "bounded unpredictability" is something like Murray Melbin's idea of night as frontier. The opening up of the night as a time of marginal activities and unpredictable events can happen only if these activities and events stay in their place within the city. There is a physical, locational analogy to night in Murray Melbin's analysis: the combat zone. But there's a defect both in the analogy and in Murray Melbin's original image: The frontier has one side that is still open, while combat zones are almost always completely bounded. Combat zones are surrounded by authoritative control of some kind. They are created, in part, by the fact that some people are making decisions that prevent certain activities from going on elsewhere.

The more-or-less deliberate placing and bounding of activities as a major feature of cities is the theme of the things I have overheard—or failed to overhear, but expected to overhear—in the discussions here. In fact, this major feature of cities has not figured very much in our discussions so far. There really has not been much discussion of power or the struggle for power in our analysis of industrial cities, or other cities. Power does include control over the location of major activities, which, in turn, determines the location of other activities. The strength and weakness of Stephan Jonas's analysis is an awareness of the importance of activity-location decisions. On the one hand, he emphasizes the importance of struggles for power among real groups, and discusses the origin of alternative programs for activity location in European cities. However, he does not give a clear idea of the technical and political limits governing these struggles. He does not indicate which of these various programs have a reasonable chance of working.

Decisions about where what kind of work will occur are crucial to other location decisions in cities. Although in the short run it is probably true that residence constrains work and determines the location of different sorts of labor, work location also constrains residence. The changing extent and character of that constraint figures implicitly in John Modell's analysis of suburbanization, but I wish it appeared more explicitly. What is the relationship between places of work and residential locations in suburban and nonsuburban Philadelphia?

We are at the tail end of the process of the proletarianization of society in Western, industrial, rich countries. Proletarianization has been going on for a long time; it results in a labor force consisting almost entirely of people who are dependent on the sale of their labor power for survival and whose means of production are controlled by others. Proletarianization, in the long run of urban history, affects who makes the location decisions about work. As we have moved toward the present over the last few hundred years, those who control capital have increasingly made the decisions about where work would be located.

Allan Pred's large job-providing organizations have increased their control over the location of work and, hence, over other activities that depend on the location of work. But the state has also acquired power. To be sure, there has been a fluctuation in the relative power of states over location decisions. For example, the state's power in seventeenth-century France did indeed fluctuate, but this fluctuation was neither a steady increase nor a steady decrease. Within Paris, the "absolute monarchy" was unable to penetrate successfully the enclosure of "le Temple," a monastically controlled free area. Neither could it control successfully industrial production in Faubourge St. Antoine, just outside the city gates. The chief of the order and the abbess who controlled these two spaces were able to hold off King Louis XIV and his agents for a long time. Why the state wanted to penetrate and regulate these spaces is itself diagnostic of what was going on in seventeenth-century Paris. Parisian merchants and tax farmers who provided the capital for the war-making efforts of the Crown insisted that these free areas provided competition for their own revenues and should not be permitted to survive within a strong monarchy.

In fact, states have often engaged in the direct, authoritarian allocation of space. Anthony Sutcliffe's question concerning the public sector and authoritative housing decisions is therefore an important one. States and governments in Europe have had a rather large role in space-allocation decisions, at least at certain times. This has also been true in the United States. For instance, something like two-thirds of the land area of Arizona belongs to the federal government; the allocation of this space depends very largely on federal decisions. What has been true for large-scale allocations has also been true for small-scale allocation decisions. Roger Lane can tell us much about the history of police street-patrolling and the struggles for definitions and control of public space.

Nevertheless, space-allocation decision in Europe have been slightly different from those in the United States. The Bédarida and Sutcliffe presentation of street life reminds me that Paris and London underwent somewhat different evolutions in regard to space allocation, especially in the period from 1780 to 1830. London and Paris look relatively similar if not identical toward the end of the eighteenth century, judging from information on street conflicts and similar events. But as we move into the nineteenth century, the differentiation between the two cities increases; something like the contrasting patterns portrayed in our essays came into being. No doubt one part of what happened in Paris and London is that the scale of production and commerce increased more rapidly in London than it did in Paris. Small commerce and small production are good ways to put people on the street and to keep them cemented in their locations. But in addition to the differences in the scale of production and commerce, the character of the control over public space—that is, policing—also accounts for some of the difference between the two cities. One of the most important changes in London in the period from the late eighteenth century to the middle of the nineteenth century was the installation of regular patrolling of what became defined in increasingly legal terms as the public space of the city. The police deliberately moved vagrants, idlers, and others from the streets of London. Although the Paris police are famous and somewhat more sensational than their British counterparts, this parallel process never went so far in Paris as it did in London.

There have been a series of fairly direct impositions of power on the allocation of space. But there have also been a series of indirect controls on the allocation of space, especially in decisions to locate different kinds of work in different parts of the city. In the United States, decisions to disperse industrial production and retail services seem to conform to fairly compelling economic advantages. But when we look at the location of administration and communication, there seems to be somewhat more choice open to those who decide on the location of the big buildings and the big organizations. In Detroit the automobile makers have been actively dispersing production throughout the metropolitan area for a good thirty years. On the other hand, they have been far less consistent on the location of the administration of the industry.

Stephan Jonas mentioned Schlumberger, the outstanding Alsatian industrialist; his decisions anchored the textile industry for a long period of time. Henry Ford III is a modern-day Schlumberger and Coleman Young is his prefect. The Renaissance Center implicitly and explicitly proposes that administration and elite services are the key to the revitalization of the central city. This idea produces a new American dilemma. The decision that has been implemented in Detroit means that there will be very few jobs for the bulk of the poor black population of Detroit.

On the other hand, Detroit has made the transition to an elite central city rather easier by tearing down a very large part of all the single-family housing near the center. Therefore, the cost of rehabilitation is not likely to figure as heavily in the reconstruction of Detroit as it has in other places that have remained continuously built up.

This case and many others like it suggest the importance of governing coalitions in determining the character and effectiveness of location policies. By analogy with studies of worker participation in industry, we might very well conclude that the external political environment is as important as the internal viability of a plan for the interior of the city—in determining whether or not it will in fact produce the desired changes. Kenneth Jackson's analysis is a sketch of the mechanisms employed in the execution of a coalition's locational policy. Lutz Niethammer's discussion provides us with a provocative hypothesis concerning the political context of another locational policy. We might even note the throwaway line at the end of Brian Berry's chapter. He argued not that greater changes would never occur but that even those small changes he was projecting would be likely to require very important alterations in the existing structure of power and government. Perhaps we can treat him as the most radical proponent of a power-based hypothesis about location decisions that we have in this discussion, rather than as a person who sees no change at all.

Having heard these things, the badaud can melt back into the crowd.

Sam Bass Warner, Jr.:

Although Chuck Tilly is surely the most stylish among us, each of us is required to be a bit different and stylish in his own way. I want to start out, as a former Midwesterner, in the most flat-footed way possible by responding to what our chairman said we attended the conference for. He said that the goals of the conference on the Dynamics of Modern Industrial Cities were to worry about whether or not urban history could say something useful. He defined "usefulness" as something that leads to urban policy. So, as I understand it, we are to consider here what we know about the dynamics of modern industrial cities. As historians we are to employ history by looking at the polarities between order and disorder. I guess we are simply trying to use the past to criticize the present and get our courage up about the future. But what is the past we have been examining? We looked at an old piece of federal housing policy, we looked at class conflicts and urban renewal, and, in Brian Berry's chapter, we actually looked at regional conflicts that have developed as the Southeast and Southwest catch up with the rest of the country. These as well as many other subjects are the sort of things talked about in the essays.

But I think the place for us to start our inquiry about whether or not urban history could be useful and what urban policy might be useful would be to ask, "What's our interest?" "What's in it for us?" "How did we all come to be here?" "Who are we?" "What is our set of concerns?" I think that looking at who we are and what our set of concerns is will tell us something about the essays that we have read and, I hope, draw the model for a policy.

Professor Niethammer has nicely characterized who we are. In his chapter, he has tried to characterize the nineteenth-century housing reformers and the international communications among them. He writes:

> The explanation for this phenomenon is to be found in the position of the reformers themselves within their own environment, where they were usually excluded from the ruling class and wealthy establishment, though some of them were quite prominent. Against the dominant class interests they were voices in the wilderness, suggesting strategies against social dangers, often personally experienced; a comfortable middle class tended to ignore them as interference with their short-term interests, and only became alarmed by intermittent waves of

epidemics and revolutionary upheavals. On the other hand, the reformers certainly did not share the working class's leftist sympathies, which they most often characterized as symptoms of a sick social body.

Their cure for economic class struggle was to introduce a new paradigm of environmental and structural control, stressing the national, spatial, and educational conditions of reproduction instead of the relations to the means of production, social biology as an essential supplement to political economy, social microorganization instead of political decision or the formation of organizations, and they tried to legitimize their position by technical expertise and empirical research.

Now I found that a rather apt description of our interests and our situation. Therefore, my first moral would be to say that our position and circumstances are not really all that different from those of the nineteenth-century housing reformers whom we look back on with some amusement. To drive the point home, I might call to your attention that twenty years ago there would not have been people with labels like "urban historian" whom you could gather in a room at a conference. While cities have perhaps been declining, urban historians have been increasing marvelously. Just as urban reformers—housing reformers—come nicely out of slums, so, perhaps, urban historians come nicely out of suburbs. As urban historians, then, our interest is clearly to be experts and for there to be things to reform. It is very clear that that is our business; there would be no purpose for us if we didn't think there were things to be reformed. But now the question is, "What is to be reformed? How are we to determine what is to be reformed?" I think the simple way to find the answer—I do not know if it is a Freudian, a Jungian, or even a Platonic device—is to ask, "What is the shadow side of our concerns?" That is, when we talk about housing, what is it we are afraid of? When we talk about unemployment, what is the opposite of our concerns?

Someone mentioned that some years ago we were much concerned with the ghetto and its turmoil, but we do not seem to be worried about that anymore. But when we were worried about it, what was it that we were worried about? Were we worried, as we say nineteenth-century housing reformers were, that the mob would come and overrun us? I think perhaps we were. Let us ask what is the problem—the danger—that people think the city has? It could be that there is a certain low temperature to our discussion, because we do not perceive any very severe dangers. The blacks have quieted down, so there is no longer a crisis. There were 200,000 people gathered in New York City recently, and they thought the problem was atomic energy. Alright, that is one possible thing, and I suppose if we ask what the flip side of that was, they had some sort of holocaust in mind. So that was their fear. Most planners in the Northeast are worried about the fiscal problems of their municipal employers, and they tend to spin out notions of unemployment and falling tax revenues and the decline in the speculative value of land. For them, these things are the "problem."

But these issues are not the major problems with which we are concerned here. Ken Jackson's people were particularly worried about a collapse in the banking system and unpaid mortgages. Their great fear was a banking collapse, and they took steps against it. I think Professor Niethammer's reference here is to an urban mob; there were lots of urban mobs in the nineteenth century and sanitation and order therefore had a very direct, concrete meaning to nineteenth-century reformers.

Professor Jonas has raised some very interesting questions, which I think might tell us the most about who we are and what we are worried abou.. I am going to compare his essay to Bédarida and Sutcliffe's because I think there's a cluster of concern here which might tell us what we are about and what concerns us. Professor Jonas reports that there is a reaction to wild growth in Europe. There's something going on in Europe that resembles what is going on here. People think growth is out of control. There is something called giantism. There's a lack of enthusiasm for and loss of faith in technology. The feeling is that Dupont is not going to bring us more wonders through chemistry, but in fact is going to poison us. Thus, there is a group of people whom we might call "greens," people who are concerned with ecology and open space and clean environments and growing things. Then there are the old Leftists, 1968 sort of people who are concerned about decentralization. These are the people who talk about local power and who are out trying to organize in the neighborhoods. In Massachusetts we have something called Fair Share, which is a descendant of that kind of group, and I believe there are groups like this in Europe. When I was in Sweden, young people had occupied a series of tenements that the municipal authorities of Stockholm were going to condemn. They were trying to save the neighborhood for "the people." There is a concern both in Europe and in the United States about the environment, there is a lack of faith in technology, and, finally, there is a sense that power needs to be decentralized. These bourgeois campaigns are signs of the wonderful elaboration and energy and enthusiasm of the bourgeoisie and how it contantly multiplies new styles, new concerns, new antitheses, if you will, out of its own society.

The discussion of the streets is supplementary here because it leads to a number of elaborations in bourgeois life in the nineteenth century—clothing, housing styles, and so on. If Bédarida and Sutcliffe keep going in their work, they will soon be concerned with family life, with how the bourgeoisie protected their children in Paris, with how Parisian techniques were different from those in London, and so forth.

This discussion doubles back on the issue of the "greens." The greens are important people, because they actually represent an attempt by the middle class to control ever larger spaces. We are not content with a nineteenth-century row house and a yard, a maid, a chauffeur, and so forth to protect this space. We have moved on in the Modell paper to a whole suburb where we wish to control our turf by using the school as a possible institution of social control. Now we need to have control over an entire region. We have got to have the whole of Cape Cod whipped into shape. We need to have public parks like the White Mountains and Yosemite. We have to have our parks properly cleaned up, policed, and protected. There is an enormous ambition here, and inventiveness on the part of the middle class for creating the kind of lifestyle it wants. The middle class is making ever-larger demands for the extension of its turf.

The decentralization issue, the growth issue, the distrust of science, it seems to me, come out of the very growth of corporate society itself. On the one hand, the corporations in which we work are nicely appointed, have all kinds of equipment, good lighting, steady pay, tenure, and regular promotions. These are the benefits, and no one knows whether we are working very hard. There are abundant water coolers; we can call home at 4 o'clock and no one will criticize. Scrooge bossism is gone in these institutions, yet individual power and autonomy are gone also. And it is that sort of an issue, it seems to me, that came up in 1968 and comes up over and over again in the issues of worker control, experiments in local control, and the seizure of and dealing with the urban economy. I would conclude from this little excursion or review that as a group we wish to be experts and reformers, and, like our predecessors, we are concerned with the elaboration and multiplication of middle-class lifestyles. We approve of them; we live them; we think they are a good thing; we think they are good for people. We are not prescribing for the poor something that we do not believe in for ourselves.

The next moral that I draw from this is that enduring problems of power distribution confront reformers at all times and confront us even more than reformers of the nineteenth century. The concentration of corporate power, the multinational corporation, the large nation-state, and the large state bureaucracy are going to be called on more and more to deal with urban problems. There are enormous aggregations of power, and we reformers are weak in the face of them. We really depend on luck to have any of our reforms come true.

Also notice that whether we are talking about housing reform or changes in lifestyles on the street or whatever, all urban events are unexpected. Everything comes as a surprise. It seems not to be possible to stand in Paris in 1800 and know what a street is going to look like in 1860. It is not possible to look at the banking crisis of 1933 and know what its urban outcome is going to be. The South and the whole Sunbelt catch-up is something we have awaited for a century, yet all of a sudden it is upon us. The unexpected, the unbounded, the surprise element should be something to which we should always attend.

Finally, it seems to me, as good bourgeois and as good experts and as good technicians, we should always be concerned with equity issues. If we notice our own behavior in the past, and the unexpected behavior of cities, it seems that we should be constantly talking about the processes whereby people do not get so badly hurt as they have in the past. When a region's economy goes sour or when a neighborhood goes down, how can we make the process less destructive to people? Brian Berry has frighteningly pointed out that to the extent that the central city goes up, the suburbs will go down; if we think that the poor were ill-treated when they had control of the largest political unit, think how they are going to be when they own all those little towns out at the edge that nobody cares about. We should not lose sight of our equity concerns; which probably are one of the nice things about the bourgeois liberal tradition.

INDEX

ABOUT THE CONTRIBUTORS

FRANÇOIS BÉDARIDA is Director of the Institute of Contemporary History and Director of Research at the National Center for Scientific Research in Paris. His works include several books concerning nineteenth- and twentieth-century British society.

BRIAN J.L. BERRY is Dean of the School of Urban and Public Affairs at Carnegie-Mellon University and formerly held the Williams Professorship of City and Regional Planning at Harvard. He has published widely, and his works include *The Human Consequences of Urbanization* and the editing of *Urban Affairs Annual Reviews, Volume 11: Urbanization and Counterurbanization.*

DAVID R. GOLDFIELD is a Visiting Professor at the Art History Institute, Stockholm University, Sweden. He is currently working on a study of housing problems in European cities on a grant from the Council of Europe. His works include *Urban America: From Downtown to No Town* (with Blaine A. Brownell), and he has contributed many articles to leading history and city planning journals. He is Associate Editor of the *Journal of Urban History.*

KENNETH T. JACKSON is Professor of History at Columbia University and is founder of the Columbia University Seminar on the City. He is a well-known author and editor in the field of urban history and is presently completing a major study of nineteenth- and twentieth-century suburbanization in the United States. Jackson serves as General Editor of *The Columbia History of Urban Life* series.

STEPHAN JONAS was born in Hungary and since 1956 has been a political refugee in France. He has served as Professor of Urbanism at the Ecole d'Architecture de Strasbourg since 1970, and as Director of the Institut d'Urbanisme at d'Aménagement Regional of the University of Strasbourg since 1975.

MARTIN T. KATZMAN is Professor of Political Economy and Environmental Sciences at the University of Texas at Dallas. His research has touched on municipal finance, ethnicity, regional development in Brazil, school desegregation, and solar energy. During the 1980-1981 academic year, he held a Guggenheim Fellowship, which supported his research on the impact of energy and natural resources on the changing distribution of America's population.

ERIC E. LAMPARD teaches at the State University of New York at Stony Brook and was formerly Professor of Economic History and Adjunct Professor of Urban and Regional Planning at the University of Wisconsin. He was a member of the Social

Science Research Council's Committee on Urbanization; the Committee on Urban Economics (CUE), Resources for the Future, Inc., and the National Research Council's Social Science Advisory Panel to the U.S. Department of Housing and Urban Development. He has published widely and significantly on the topic of urbanization.

SEYMOUR J. MANDELBAUM is Chairman of the Graduate Group in City and Regional Planning and a member of the Department of History and the Annenberg School of Communications at the University of Pennsylvania. His publications include *Boss Tweed's New York, Community and Communications,* and "Urban Pasts and Urban Policies," which appeared in the August 1980 issue of the *Journal of Urban History.*

PETER A. MARCUSE is Professor of Urban Planning at the Graduate School of Architecture and Planning at Columbia University. He was a member of the Waterbury, Connecticut, City Planning Commission, as well as President of the Los Angeles City Planning Commission. His numerous publications include "Housing in Early City Planning" in the February 1980 issue of the *Journal of Urban History.*

JOHN MODELL is Professor of History at the University of Minnesota and Research Associate at the Philadelphia Social History Project, University of Pennsylvania. He is author of *The Economics and Politics of Racial Accomodation: The Japanese of Los Angeles 1900-1942* and has published many articles and contributions to books dealing with family, social, and urban history.

LUTZ NIETHAMMER teaches at the University of Essen, Federal Republic of Germany. His writings have offered insight into the recent history of his native country and he is a leading practitioner of oral history in Germany.

GIORGIO PICCINATO is Professor of Planning Theories at the Instituto Universitario di Architettura of Venice. His books include *The Territorial City: A Didactic Experiment, Contemporary French Architecture*; and *Urban Construction: Germany, 1871-1914.*

CHRISTINE M. ROSEN received her doctorate from the Department of History at Harvard University and has taught in the Department of City and Regional Planning and the Department of History at the University of California at Berkeley. At its 1980 Annual Meeting, she addressed the American Historical Association on "Walking City into Industrial Metropolis: The Problems and Process of Spatial Change in Chicago After the Great Fire of 1871."

JOHN B. SHARPLESS teaches at the University of Wisconsin in Madison. His publications include *City Growth in the United States, England, and Wales, 1820-1861* and, with Sam Bass Warner, Jr., "Urban History" in the *American Behavioral Scientist* 21 (1977).

BRUCE M. STAVE is Professor of History and Director of the Center for Oral History at the University of Connecticut. He was the organizer and coordinator of the Conference on the Dynamics of Modern Industrial Cities. His publications include *The*

Making of Urban History and *Socialism and the Cities.* He is Associate Editor of the *Journal of Urban History.*

ANTHONY R. SUTCLIFFE is Reader in Urban History in the Department of Economic and Social History, University of Sheffield, England. He is editor of *The Rise of Modern Urban Planning, 1800-1914* and author of *The Autumn of Central Paris: The Defeat of Town Planning, 1850-1970,* among other works.

CHARLES TILLY is Professor of Sociology and History and Director of the Center for Research and Social Organization at the University of Michigan. His books include *The Vendée, An Urban World, The Rebellious Century* (with Louise and Richard Tilly) and *From Mobilization to Revolution.*

SAM BASS WARNER, Jr., is William Edwards Huntington Professor of History and Social Science at Boston University. His work has served as a formative influence on the development of U.S. urban history and includes the following books: *Streetcar Suburbs, The Private City, The Urban Wilderness, The Way We Really Live,* and *Measurements for Social History* (with Sylvia Fleisch).